mestizo genomics

mestizo genomics

Race Mixture, Nation, and Science in Latin America

Peter Wade, Carlos López Beltrán,
Eduardo Restrepo, *and*
Ricardo Ventura Santos,
editors

Duke University Press | *Durham and London* | 2014

Printed in the United States of America on acid-free paper ∞
Designed by Barbara E. Williams
Typeset in Officina Sans and Quadraat by BW&A Books, Inc.

LIBRARY OF CONGRESS CATALOGING-IN-PUBLICATION DATA
Mestizo genomics : race mixture, nation, and science in Latin America
/ Peter Wade, Carlos López Beltrán, Eduardo Restrepo, and Ricardo
Ventura Santos, editors.
pages cm
Includes bibliographical references and index.
ISBN 978-0-8223-5648-6 (cloth : alk. paper)
ISBN 978-0-8223-5659-2 (pbk. : alk. paper)
1. Mestizos—Latin America. 2. Latin America—Race relations—
History. 3. Genomics—Social aspects—Latin America. I. Wade,
Peter, 1957– II. López Beltrán, Carlos, 1957– III. Restrepo, Eduardo.
IV. Santos, Ricardo Ventura, 1964–
F1419.A1M48 2014

305.80098—dc23   2013026441

contents

preface

The roots underlying the project that gave rise to this book—in a way perhaps befitting the subject matter—go back a long way. As a postdoctoral researcher doing fieldwork in Colombia in the mid-1980s, Peter Wade first encountered Eduardo Restrepo, who was an undergraduate at the time. The meeting was the beginning of a long series of encounters over the next few decades. Not long after, while doing his doctorate in the United Kingdom, Carlos López Beltrán got to know Peter Wade in Cambridge, via a mutual Mexican friend, Alfonso Martín del Campo. After a long hiatus, their acquaintance was renewed at a conference on populations of African origin held in Veracruz in 2008, at a time when López Beltrán, along with his colleague Francisco Vergara Silva, had already been writing about the Mexican genome project. In the meantime, Ricardo Ventura Santos had sent Wade a copy of the article he coauthored and published in *Critique of Anthropology* (2004) on race and genomics in Brazil. So when Wade began to tinker with the idea of a project on genomics and race in Latin America, the infrastructure of the collaborations was already in place, transnational in scope and crossing the disciplinary boundaries of social anthropology, cultural studies, the history and philosophy of science, and biological anthropology.

Luckily, our timing was right and the project met with favorable reactions from the Economic and Social Research Council, United Kingdom, which agreed to fund the research for eighteen months (grant RES-062-23-1914). The funding included salaries for three postdoctoral researchers, to be based at the University of Manchester (María Fernanda Olarte Sierra, Michael Kent, and Vivette García Deister), and three part-time research assistants, to be hired in each of the three Latin American countries (Adriana Díaz del Castillo, Mariana Rios Sandoval, and Verlan Valle Gaspar Neto). We also had money to fund a number of project workshops and we were very glad to have the Mexican biologist Francisco Vergara Silva as a constant companion in these meetings.

After an initial three months of preparation in Manchester, fieldwork was carried out in Latin America for nine months, mainly by the postdocs and, as it turned out, the local research assistants. This work focused on the geneticists and their laboratories and involved participant observation in the labs, interviews, and analysis of written materials. As described in the appendix, methods varied a little between countries: the focus in Mexico on the national medical genetics institute, INMEGEN, allowed García Deister a particularly in-depth relationship with a small number of scientists and technicians there. In Colombia, the diversity of genomics research meant the net was spread a little wider by Olarte Sierra working closely with Díaz del Castillo, while in Brazil, Kent found himself traveling the length of the country to encompass some of the great variety of genomics research there. In all cases, and as a result of the ethnographic methods employed, our researchers ended up concentrating on a small number of labs and scientists, with a focus on the way these human populational genomics projects operated in practice, the categories and methods they used to proceed, the reasons they took the shape they did, and how their results circulated, including domains beyond the science labs.

The regular workshops, held every three months during fieldwork and once after it ended, were fundamental to the working of the project. One of the reviewers of the manuscript of this book was interested in how the transnational exchanges inside the project's research team influenced the ideas that appear here. In fact, the transnational composition of the team was less significant than the comparative dimensions of the data that were emerging. It was not as if each team member brought a specific national approach to understanding the issues. One might have thought that a concept such as race could be a bone of contention for scholars coming out of British, Dutch, Brazilian, Colombian, and Mexican academies, but, in fact, as scholars of Latin America—most of whom had had transnational training experiences—we operated pretty much with a common understanding of the concept and its vagaries in the Latin American context. More generative was the experience of seeing how some aspects of genomics, especially in more public modes, were inflected by the national contexts in which it operated: for example, the emphasis on regional variety in Colombia, the nationalist rhetoric at work in Mexican biomedicine, or the emphasis by some Brazilian geneticists on the nonexistence of biological race and the illegitimacy of race (biological or social) as a basis for any kind of public policy, such as affirmative action. Yet we were also struck by the variety within each country and by the similarities between them—the use of genetic data to reinforce the gendered narratives of the origins of the nation in the sexual encounter between European men and Amerindian or African women was a common thread, for example.

The different disciplinary perspectives that team members brought to bear was also a vital part of the workshops. As it happened, the genealogy of genetics as a discipline was of interest to various people, whether historians, social anthropologists, biologists, or biological anthropologists by training. This historical perspective was a constant reminder of the dangers of presentism in studying contemporary genomics and of how much of what we were looking at, despite the new technologies being employed, had deep roots in the past. On the other hand, one aspect that had little effect on the internal dynamics of our research team was differing stances with regard to matters of politics and policy, such as the value of affirmative action in higher education in Brazil. Although some observers found it odd—or even suspect—to see people with different political views collaborating together, we found this relatively easy to negotiate inside the team.

The workshops all included open sessions to which other academics, students, the press, and the general public were invited, as a way of disseminating our research. In Latin America, these sessions were well attended. Included in the invitation were some of the geneticists with whom we were working. Relations with these scientists were potentially a thorny issue. We were interested in whether and how categories such as race entered into their work. Most of the scientists rejected race as a valid biological category and might regard any implication that race was somehow still at work in their research as erroneous and even offensive. Researchers like us who tried to reveal underlying processes of racialization could seem patronizing and arrogant, as if the scientists needed these researchers to show them things they were not aware of themselves. In some instances, it has proven a difficult path to tread in the analysis and writing stages and on a couple of occasions some geneticists reacted negatively to our arguments (or what they understood our arguments to imply). While intent on understanding the science in its own terms and context, we also argue that some assumptions that are built into the normal practice of genomics can reinscribe—in altered form—concepts and categories that look like race, especially to the nongeneticist. During the fieldwork with the geneticists, the fact that we took their projects and their practices seriously and spent time delving into them made it easier to establish a productive dialogue with them, as various chapters in this book demonstrate (see, for example, chapter 5; see also the exchange in Bortolini 2012; Kent and Santos 2012a, 2012b). Even when the scientists' reactions were negative there was a process of dialogue, which caused us to revise several passages in the book.

The second phase of the project is only marginally represented in this book. It began in August 2011 with eighteen months of funding from the Leverhulme Trust (grant RPG-044) and focused on "public engagement with genomic research and race in Latin America," building on the first phase, but with a greater emphasis on how scientific knowledge about population genomics circulates through scientific and nonscientific public spheres and how diverse publics engage with this knowledge. Some changes in personnel took place for this second phase, which is reflected in the participation in chapter 2 of this volume of Ernesto Schwartz-Marín (postdoctoral researcher for Colombia) and Roosbelinda Cárdenas (local research assistant for Colombia).

Finally, it is necessary to mention the publication of a Spanish-language version of this book (the text of which is not exactly the same, as some very minor revisions were done to the English-language version after the translation work had been completed—for which thanks to Sonia Serna). It has been important to our team to publish the results of our work in Latin America, and Carlos López Beltrán, Eduardo Restrepo, and Ricardo Ventura Santos all worked hard to create a copublishing collaboration between the Fondo de Cultura Económica and UAM Cuajimalpa (Mexico), Editorial Universidad del Cauca (Colombia), and Editora Fiocruz (Brazil). Vivette García Deister has played a leading role in coordinating the translation and editing of the Spanish-language book.

Acknowledgments

We are grateful to the Economic and Social Research Council and to the Leverhulme Trust for their support, outlined in the preface. In multiple ways, we all owe a personal debt of thanks to families and partners who supported each of us during the research—in no particular order: Sue Wade, Paty Costas, Mónica Benítez, Juan Antonio Cruz, and Manuel Cruz (also a constant companion and "product" of this research project).

We would like to thank the following people and institutions.

In the United Kingdom: Liverpool Microarray Facility (University of Liverpool) for the guided tour of their facility they gave some of us; Jeanette Edwards and Penny Harvey (University of Manchester) for their help and support; John Pickstone (University of Manchester) for his interest in the project; and Susan Lindee, Jenny Reardon, Gisli Pálsson, Amade M'charek, and Andrew Smart for their participation in a project conference in Manchester in July 2011. In Brazil: Maria Cátira Bortolini, Francisco Salzano, Sidney Santos, Ândrea Ribeiro dos Santos, João Guerreiro, Sérgio Pena, Tábita Hünemeier, Rita Marrero, Vanessa Paixão Côrtes, Caio Cerqueira, Eduardo Amorim, Pablo Abdon, Elzemar Ribeiro, Eliseu Carvalho, and many other geneticists who generously opened their doors and donated their time at the Federal Universities of Rio Grande do Sul, Pará, and Minas Gerais, as well as the State University of Rio de Janeiro. In addition, we thank Antonio Carlos Souza Lima, Claudia Fonseca, Glaucia Silva, Jane Beltrão, Marcos Chor Maio, Peter Fry, Ruben Oliven, Penha Dubois, and Thereza Menezes.

In Colombia: María Fernanda Olarte Sierra and Adriana Díaz del Castillo are especially indebted to William Usaquén, Angela Alonso, Andrea Casas, Leonardo Eljach, Verónica Rocha, Madelyn Rojas, Wilson Rojas, Vanessa Sarmiento, and Blanca Shroeder for generously opening the doors of their lab and their daily lives to them; for patiently answering their questions and providing them with thorough explanations; but especially for respecting their work and their analysis of them, engaging with them in fruitful discussions as fellow researchers, which made their chapter what it is: a back and forth of ideas and arguments. In addition, they thank Stuart Blume, Tania Pérez-Bustos, and Andrew Smart.

Eduardo Restrepo, Ernesto Schwartz-Marín, and Roosbelinda Cárdenas wish to thank Gabriel Bedoya, William Arias, and Claudia Jaramillo at Genmol in Universidad de Antioquia for their willingness to endure repeated visits and answer endless questions. Andrés Ruiz Linares, both in Colombia and in the United Kingdom, extended his kindness and generosity to support their endeavors. William Usaquén at Universidad Nacional kindly provided them with long interviews in his laboratory. At the Universidad Javeriana's Instituto de Genómica Humana, they wish to thank Jaime Bernal, Alberto Gómez, and Ignacio Zarante. At the offices of the national police in Bogotá, they wish to thank the teaching staff and directors of the School of Criminal Investigations. At the Instituto Nacional de Medicina Legal y Ciencias Forenses (INMLCF), they are particularly indebted to Patricia Gaviria, Aida Galindo, Esperanza Jiménez, and Manuel Paredes for their generosity. Finally, their work would not have been possible without the generosity and hard work of everyone at the Universidad Javeriana's Instituto de Estudios Culturales y Sociales (Pensar). The institute's staff and especially the director, Alberto Múnera, SJ, were an unwavering source of support for the project. They wish to extend special thanks to Silvia Bohórquez, Angélica Arias, Maria Fernanda Sañudo, Deimy Veloza, Gloria Chacón, and Marlén Garzón.

In Mexico: Víctor Acuña, Rubén Lisker, Rafael Montiel, Andrés Moreno, Karla Sandoval, and especially the researchers, technicians, and staff at INMEGEN (Irma Silva-Zolezzi, Juan Carlos Fernández, Fabiola Morales, Leticia Sebastián, Alejandra Contreras, Santiago March, Alejandro Rodríguez, José Bedolla, and Enrique Hernández Lemus), who kindly allowed Vivette García Deister to intrude in their busy work days and even went out of their way to accommodate her curiosity. Also the Instituto de Investigaciones Filosóficas, at UNAM, for administrative work and support during the project meetings, particularly Guillermo Hurtado, Amelia Rodríguez, and Amado Luna. Francisco Vergara Silva accompanied us all the way, while the members of the Seminario de Genómica Crítica at UNAM provided an excellent discussion arena. We received constructive comments from Rasmus Winther, Edna Suárez, and Ana Barahona.

At Duke University Press: the anonymous reviewers for their valuable comments on earlier versions of the book, Gisela Fosado for her supportive work as a commissioning editor, and Liz Smith for her help with the production process.

Genomics, Race Mixture, and Nation in Latin America

Peter Wade, Carlos López Beltrán, Eduardo Restrepo, and Ricardo Ventura Santos

This book presents findings from an interdisciplinary project involving three research teams working in Brazil, Colombia, and Mexico. Collaborating closely, the teams carried out in-depth research in a small number of genetics laboratories in these countries, while also drawing on local histories of physical anthropological and biomedical research into human biological diversity.

Laboratories in Brazil, Colombia, and Mexico have been mapping the genomes of local populations, with the objectives of locating the genetic basis of diseases and of tracing population histories. Geneticists are frequently concerned to calculate the European, African, and Amerindian genetic ancestry of these populations or to compare them to samples of European or Amerindian populations. In the process, scientists sometimes link their findings explicitly to questions of national identity, racial-ethnic or population difference, and (anti)racism, stimulating public debate and sometimes engaging in the definition of public policies.

The chapters in this book explore how the concepts of race, ethnicity, nation, and gender enter into these scientific endeavors and whether these concepts are reproduced, challenged, or reformulated in the process. Our work links current research in genetics to recent changes in the three countries, which in the last two decades have moved toward official multiculturalism, as have many countries in Latin America. The way genetics creates new imagined genetic communities, which may take forms that, to observers outside the genetic field including experts from other areas (anthropologists, sociologists, historians, etc.) and laypeople, might appear to have racialized and national dimensions, has implications not only for changing conceptions of race, ethnicity, and nation, but also for citizenship and social inclusion and exclusion.

The growing literature on race, identity, and genomics focuses mainly on the United States and Europe. Latin America, with its national identities based on *mestizaje* or *mestiçagem* (roughly translatable as "race mixture" in Spanish and Portuguese, respectively), presents a fascinating but little-explored counterpart (see, however, Gibbon, Santos and Sans 2011; López Beltrán 2011). Our project team, comprising eight Latin Americans based in Latin American countries and two Europeans based in Europe—all with extensive experience of the Latin American context—was well placed to approach this literature from a different angle. In Latin America, clear categories of racial identification have been blurred by centuries of biological and cultural mixture, and ideas of race are often less socially salient than in the United States or appear in a very cultural form, in which biology and naturalization are often seen as less important. Critical race studies of Latin America have challenged the myths of "racial democracy" that have sometimes been erected on the basis of these characteristics, also showing that Latin America is very diverse with respect to ideas and practices around race (Maio and Santos 2010; Restrepo 2012; Wade 2010). Still, a view from south of the border tends to take race less for granted as a normal part of the social and political landscape and thus to question in detail exactly what is being reproduced when concepts of race, ethnicity, and nation become entwined in genetic research.[1]

Our research indicates that, despite the fact that most geneticists in Latin America actively deny the validity of race as a biological category, genetic science might produce knowledge and interpretations that, while they appear nonracial to genetic experts, might look a lot like race to the nonexpert in genetics. This occurs in social contexts in which race has a particularly contested presence and definition to begin with (as we show later on). Even when scientists explicitly deny the association between ancestry and race, the way genetic knowledge reaches society at large can give unintended but public salience to a notion of race based on ancestry—to be specific, biogeographical ancestries of continental scale (African, European, Amerindian)—in a context in which *raza* or *raça* (race) can evoke many different ideas of ancestry, appearance, culture, class, region, and nation. In emphasizing biogeographical ancestry measured through the use of selected DNA markers, genetics rejects older notions of race as biocultural types of human beings, yet it can be interpreted and understood (outside the genetic field) as reinforcing commonsense understandings of human diversity as divided up into continent-shaped groups.

Ancestry is not the only way in which ideas about race may recirculate. Genetic science in Latin America frequently evokes the nation and, insofar as the nation has long been a key vehicle carrying the idea of race in the region, and race has likewise been a central category in Latin American nation building, this evocation in genetic idiom can also entrain racialized meanings. In addition, a gendered discourse (mestizaje and the nation as originating in sexual relations between European men and indigenous or African women) is particularly powerful in Latin American genomics. While genetics tends to highlight ancestry, this is mediated through ideas about nation and gender in ways that reiterate the diverse meanings of raza in Latin America.

At the same time, commonsense notions of race are transformed as well as being reinforced. Genetic reckonings of ancestry involve abstraction, metrification, differentiation, and multiplication—processes discussed in detail in the conclusion—which lead to contradictory effects of stabilization and destabilization, of fixing and unfixing, in which categories of race, ethnicity, and nation reappear to nongeneticists' eyes in a genomic idiom, which however simultaneously contests that appearance. The Latin American material brings into sharp focus, first, the way genetics can operate to biologize and naturalize commonsense and vague ideas about race, while also multiplying the diverse meanings of race; and, second, the way genomics can do this via concepts of nation and gender.

As we will see throughout this book, the categories used in genetic research are far from being neutral technical devices—as is the case with many scientific categories. Instead, they are natural-cultural objects that circulate through both scientific and nonscientific realms, blurring the boundaries between these realms, acquiring many different meanings and being subject to different readings.

Human Diversity Research and Race Studies

The History of Researching Human Diversity

Interest in human diversity has a very long history, in which what we would now call the cultural and the biological have not always been separated out in the clear fashion to which Western thinkers are now accustomed (Hodgen 1964; Jardine, Secord, and Spary 1996; López Beltrán 2007; Thomson 2011; Wade 2002b). In the nineteenth century, when biology and physical anthropology began to emerge as the distinct fields of enquiry they were to become, the division of humans into categories according to their physical characteristics—their biology—became increasingly established as a specific endeavor (Lindee and Santos 2012; Marks 1995; Spencer 1997; Stocking 1982, 1988). Biology, however, was often regarded as interwoven not only with environment, but with behavior and habit too.

The idea of race emerged as early as the thirteenth or fourteenth century to refer to lineage, breed, or stock in animals and humans (Banton 1987; Stolcke 1994).[2] It became entwined in notions of "purity of blood" (limpieza de sangre) and religious affiliation, especially in Iberian encounters between Christians,

Jews, and Muslims (Hering Torres 2003; Poole 1999; Sicroff 1985; Zuñiga 1999). During the discovery and conquest of the Americas, this idea of lineage and purity of blood became one way (among others) of thinking about differences between the key categories that emerged in these colonial encounters—blacks/Africans, whites/Europeans, Indians/Native Americans (as well as Asians and other non-Europeans) (Martínez 2008; Rappaport 2012; Villella 2011). Differences in perceived appearance as well as behavior were naturalized in ideas about heredity.

During the eighteenth and nineteenth centuries, the idea of race developed as the key conceptual category with which to classify humans into "types." By the nineteenth century, with the development of biology and physical anthropology, these types were conceived as physically distinct entities, even separate species, and were ranked in a hierarchy of biological and cultural value (Banton 1987; Restrepo 2012: 153–173; Smedley 1993; Stepan 1982; Stocking 1982). This idea of race—often called "scientific racism"—remained influential in the early decades of the twentieth century. According to many narratives, the concept began to be dismantled from about the 1920s, as scientific evidence mounted against it as a workable classificatory device. It was then consigned to the dustbin of science in the wake of Nazism, with a vital role being played by the postwar statements on race produced by UNESCO in 1950 and 1951, about the lack of scientific evidence for a biological hierarchy of racial types (Barkan 1992; Haraway 1989; Reardon 2005; Stepan 1982).

Race then remained as a social construct, an idea that people, not scientists, used to categorize themselves and others, perhaps referring more to culture than biology, perhaps avoiding the term itself, but still generally using classic racial phenotypical cues to make categorizations, typically essentialist in character and typically referring to familiar racialized categories (black, white, Asian, African, European, mixed-race, Native American or *indio*, etc.). It is this combination of reference to physical appearance, heredity, essences, culture, and the specific categories of people that emerged in colonial histories that we, in this book, take as defining "race."

In fact, the use of the concept of race as a way of thinking about human diversity, if not hierarchy, did not disappear in the life sciences in the period following World War II (Maio and Santos 1996; Reardon 2005; Reynolds and Lieberman 1996). Many life scientists amassed biological, including genetic, evidence to indicate that humans could not be biologically partitioned into separate entities called races (Brown and Armelagos 2001; Montagu 1942). Humans, as a young species in evolutionary terms, did not evolve into clearly distinct types in different geographical and demographic niches; in addition, they constantly moved and interacted, with increasing frequency over time. Still, life scientists worked with and developed the idea of populations, which could be understood in terms of demography and/or genetics, and which were dynamic and not clearly bounded. In this view, populations were still distinguishable in terms of frequencies of certain traits and, in global and evolutionary terms, could still be broadly differentiated into continental-scale populations that looked rather like the older races or evoked race-like notions, now in terms of genetic frequencies (M'charek 2005a). Influential geneticist Cavalli-Sforza and his colleague Bodmer even argued for "a genetic definition of race"—one more accurate than everyday concepts of race—in a book published in 1971 and republished in 1999 (Reardon 2005: 54, 70), even though Cavalli-Sforza also argued that clusters of populations could not be identified with races (Cavalli-Sforza, Menozzi, and Piazza 1994: 19).

With advances in genetics, this picture became increasingly complex. On the one hand, the fact that humans all share 99.9 percent of their genomes acquired iconic status. On the other hand, new technologies gave scientists increasing powers to explore that 0.1 percent of difference, which by definition caused much of the evident physical diversity of humans. Long-standing interest in mapping this diversity expressed itself anew in the Human Genome Diversity Project, begun in 1991, which, despite its checkered career, created a database that is widely used today. Other global initiatives include the Polymorphism Discovery Resource (1998), the International HapMap Project (2002), the Genographic Project (2005), and the 1000 Genomes Project (2008).

The aims of mapping diversity revolve around (a) understanding processes of human evolution and global migrations—which may include a kind of rescue genomics, seeking to sample "isolated populations" before they disappear or lose their supposed genetic particularities (Abu El-Haj 2012; Marks 2001; Reardon 2001; Santos 2002; TallBear 2007); (b) improving human well-being, by locating genetic components of disorders, using techniques that compare different populations or that control for genetic differences related to geography; and (c) developing databases for the forensic identification of individuals.

The aims related to medical genomics are very powerful. More recent debates about race in genetics revolve centrally around health. The issue is whether variations in the incidence of disease and in human drug responses relate in significant ways to genetic differences that may be characterized as racial, ethnic, or more neutrally in terms of continental-scale "biogeographical ancestry" (with the latter usually breaking down into categories such as African, European, Asian, Amerindian) (Abu El-Haj 2007; Burchard et al. 2003; Cooper, Kaufman, and Ward 2003; Fujimura, Duster, and Rajagopalan 2008;

Fullwiley 2007a, 2008; Koenig, Lee, and Richardson 2008; Nash 2012b; Skinner 2006, 2007; Whitmarsh and Jones 2010).

How Race Can Enter Genomic Human Diversity Research

Many geneticists reject the concept of race as a meaningful biological category, although some do not (Bliss 2009b; Burchard et al. 2003). Several leading Latin American geneticists have campaigned explicitly against the concept of race (Lisker 1981; Pena 2005, 2006, 2008), and all of the geneticists we worked with resisted the suggestion that their work somehow reproduced race—a suggestion that they often understood to be the accusation that they were reproducing old-style scientific racism. This resistance produced important exchanges between us and the geneticists, which helped us clarify our arguments.

There are several reasons why the concept of race or something that appears to nonexperts in genetics to invoke categories that look very like racial ones—if not the term itself—seems to derive from genomics research. First, research into human genetic diversity may employ sampling strategies that seem to evoke familiar race-like categories, while referring to "populations" (TallBear 2007). Thus the HapMap includes samples of Yoruba people from Ibadan and people from Utah with northern European ancestry and, despite HapMap cautions about generalizing, these samples are often used as proxies for African and European ancestry (Bliss 2009a). The HapMap organizers deny any connection with race, yet sampling "the Yoruba" ends up being a way to sample "Africans" and to specify "African ancestry" (Reardon 2008: 314). Or, as we found, DNA extracted from Zapotec individuals in Mexico is used as a proxy for Amerindian ancestry. The use of AIMs (ancestry informative markers) is "designed to bring about a correspondence of familiar ideas of race and supposed socially neutral DNA": populations of Africans, Europeans, and Native Americans are sampled and then used as "putatively pure reference populations" to define the genetic ancestry of "admixed" populations, as geneticists generally call them (Fullwiley 2008: 695).

For the geneticists, AIMs are selected genetic markers that can help to identify geographical ancestry; AIMs do not comprise the genetic profile of an entire population and they may bear no relation to phenotypical expression, being located in the noncoding sections of the DNA.[3] For geneticists, the use of these isolated markers is far removed from early twentieth-century concepts of races as clearly defined biological units, with associated behavioral characteristics (Abu El-Haj 2012). Yet for the nongeneticist observer, the use of reference populations to identify African, European, and Native American ancestry in admixed populations almost inevitably reinscribes, not race in a simple sense, but racialized concepts of human difference, as it reiterates these familiar categories, with the apparent underlying implication that they are biologically distinguishable as populations.

Second, specific diseases or conditions may be associated with particular populations or with particular ancestries, in ways that again evoke familiar race-like categories and seem to give them a genetic basis. Thus Mexican Americans or Mexican ancestry may be associated with type 2 diabetes (Montoya 2011); African ancestry may be associated with the frequency of asthma (Fullwiley 2008); and medicines may be marketed to Americans of African ancestry (Kahn 2008). Related to this issue is the question of whether, and exactly how, racial and ethnic labels should be used in medical research and clinical practice. Differences—and especially inequalities—in health outcomes that are structured by racial and ethnic identities can be measured by using such labels, but they may also entrench differences and even invite geneticized explanations for health differences, which may be more social than genetic in origin (Braun et al. 2007; Ellison et al. 2007; Kahn 2005; Kaplan and Bennett 2003; Koenig, Lee, and Richardson 2008; Whitmarsh and Jones 2010). The use of racial and ethnic labels may become quite standardized: in her study of a U.S. genetics lab, Fullwiley (2007a: 4) found that, because it worked as a classifier in many different U.S. contexts, race functioned as an institutionalized way to organize and interpret data (Epstein 2007). Genomics does not, however, necessarily associate disease with race-like categories; it may contest such associations, as is the case for sickle-cell anemia in Brazil, where state screening programs, based on genetic evidence, test all members of state populations, despite strong discursive links between the condition and "black people" in Brazil (Fry 2005b; see also Pena 2005). This is an important indication that genomics does not simply reproduce but may also destabilize racialized categories (see also Pena 2008).

Third is the technique of controlling for "population stratification." A medical genomics project that is seeking a genetic variant associated with a given condition generally compares diseased cases with healthy controls. If the cases happen to possess various genetic variants more frequently than the controls, then a number of genetic variants may appear to be associated with the disease, some or all of which may have nothing to do with the disease, being the product of other demographic or evolutionary processes. One possible difference between cases and controls is caused by the genetic ancestry of populations: populations have different genetic profiles by virtue of their ancestral geographical location. Thus it is important to make sure that your cases and controls are matched in terms of ancestry: if you compare African cases to European controls you will find lots of genetic differences, without being able to tell which are simple accidents of geographical ancestry and which are actually linked to the disorder in question (Fujimura and Rajagopalan 2011). Basic matching is generally done by asking sampled individuals to self-identify in ethnic or racial terms.

Matching may need to be refined further: because of historical admixture, any sample of "Mexican" cases or controls is likely to include individuals with very varied mixtures of ancestries, many of whom might identify as mestizos. But by genotyping each individual to quantify biogeographical ancestries (in this case, mainly Amerindian and European), it is possible to statistically control for admixture, so that cases and controls are finely matched (Choudhry et al. 2006). A genetic trait that is linked to, say, asthma will thus hopefully stand out, independent of other traits that happen to be associated with a given ancestry. Of course, once you have good evidence that a given genetic trait is linked to a disease, you may find that this trait is more prevalent among certain biogeographical populations and ancestries than among others.

The three preceding points show that biogeographical ancestry appears again and again in relation to populations conceived as different because of their evolutionary and demographic histories in diverse geographical locations. This evokes race-like categories through "genome geography" (Fujimura and Rajagopalan 2011), even when the scientists in question deny the biological validity of race and see ancestral populations defined by specific sets of markers as quite unlike races. It is important to emphasize that these are race-like or racialized categories, rather than simple reiterations of early twentieth-century racial types. The categories referred to are populations (which may be quite specific) and ancestries inferred from selected populations. Yet, to the eye of the nongeneticist, larger, race-like categories, such as African, European, and Amerindian, constantly emerge from these more specific references.

A fourth way in which racialized thinking appears in genomics is via the idea of the nation. Race and nation have long been linked in the idea that nations are biologically distinctive in ways linked to racialized classifications, such that Britishness or Englishness might be conceived in terms of whiteness or Brazilianness seen in terms of having mixed ancestry (Anthias and Yuval-Davis 1992; Appelbaum, Macpherson, and Rosemblatt 2003b; Foucault et al. 2003). To the extent that genomic research attempts to create a national genomic science, a national biobank, or a map of national genetic diversity, there is the possibility that the idea of the nation will be given a genetic connotation, albeit an imagined one, which may reinforce racialized meanings (Benjamin

2009; Gibbon, Santos, and Sans 2011; Hartigan 2013a; López Beltrán 2011; Maio and Santos 2010; Nash 2012c; Pálsson 2007: ch. 4 and 5; Rabinow 1999; Taussig 2009).

Race, Genomics, and Society

These specific ways in which race-like categories appear to the nonexpert eye in genomic research need to be seen in the context of the impact of genomics on society more generally. Existing literature on the social changes wrought by the advent of the "new (human) genetics" presents us with a number of possibilities.[4] "Geneticization" refers to "an ongoing process by which differences between individuals are reduced to their DNA codes" (Lippman 1991: 19) and, more generally, implies a shift toward conceiving of belonging and identity in terms of genetic traits. This appeal to genetic determinism can be paradoxically combined with choice and self-fashioning, for example, through "recreational" ancestry testing (Bolnick et al. 2007; Comaroff and Comaroff 2009: 40). "Gene fetishism" involves reifying and attributing powerful agency to the gene, and also turning it into a cultural icon (Haraway 1997; Nelkin and Lindee 1995). "Biosociality" suggests that people will increasingly create social relations around perceived biological ties, such as shared genetic disorders (Gibbon and Novas 2007; Rabinow 1992; Taussig, Rapp, and Heath 2003). "Biological citizenship" refers to the use of biological traits to define belonging and entitlement in a nation-state (Rose and Novas 2005); in this, biomedical classifications are increasingly important and may entail genetic considerations (Heath, Rapp, and Taussig 2007).[5]

These concepts are varied, but they share the idea that the biological and the genetic are an ever greater part of social life. While this is undoubtedly true in a general sense, some literature also shows that this is an uneven process, with some, perhaps many, sectors of different societies having only superficial contact with genetics, while it is not clear that their ideas about identity and belonging have become geneticized or even biologized (Edwards and Salazar 2009; Hedgecoe 1998; Wade 2007b). To the extent that geneticization has occurred, there is also no simple sense of the consequences of the greater social presence of genetics, although some have feared that there is a shift in society toward greater genetic determinism and reductionism (Lewontin, Rose, and Kamin 1984), even if this goes against current genetic science, which tends to be less deterministic than before (Keller 1995; Pálsson 2007: 44–49).

Discussions about the concept of race are significant in relation to these ideas. Concerns about the reappearance of racialized thinking in genomics are linked to worries about processes of geneticization and genetic determinism.

The fact that race—usually understood to be deterministic—seems to be re-iterated in some form in genomics encourages fears about this.[6] However, recent work on genetics and society suggests that such a characterization of race and genomics is overly simple, in part because geneticization is not so-cially pervasive as a way of thinking about belonging and identity. In addition, just as geneticization does not necessarily lead to greater determinism, partly because apparently rigid separations between nature and culture are not as clear-cut as they appear and are blurred still further precisely by geneticization itself, so the concept of race, as a natural-cultural assemblage, is not simply about biological determinations (it is, and always has been, also about envi-ronmental determinations and cultural habits).[7] Thus the reappearance of race need not only lead to greater determinisms, although this may be one aspect of the story (Condit 1999; Condit et al. 2004; Nelson 2008b; Roberts 2010; Wade 2002b).

Race, Gender, and Human Diversity Research in Latin America

The preceding discussion sets one key frame for the findings from our proj-ect. Latin America has attracted interest for some time, due to the presence of indigenous peoples, who can provide data for understanding microevolution-ary processes and human migrations (Salzano and Callegari-Jacques 1988), and due to the presence of genetically admixed populations, which are seen as useful for tracing interesting genetic variants, untangling genetic and en-vironmental causes of complex disorders, and making inferences about pop-ulation migrations (Burchard et al. 2005; Chakraborty and Weiss 1988; Dar-vasi and Shifman 2005; Salzano and Bortolini 2002; Sans 2000). The impact of genetic medical and human diversity research on Latin America has been increasing over the last twenty years or so.

One of our conclusions is that, due to the complex routes through which science reaches the public at large, genomics in Latin America can provide a genetic language for thinking about something recognizable in the wider so-ciety as race—but it does not do this in a simple, unidirectional way and indeed it complicates race in the process. The reinscription of racialized concepts is particularly apparent in societies where a public discourse of race has either been historically marginalized since the demise of eugenics (as in Mexico) or has been subsumed into discourses about mixture and the transcendence of racial difference (as in Brazil); and in societies where a biological language of race—although by no means absent, especially insofar as phenotypical ap-pearance can be important in making racialized judgments—has historically been less evident than in the United States or Europe. (We expand on this in the next section.)

However, our material also shows how genomics provides a different way of thinking about race so that it appears in a new molecular and bioinformatic mode. As noted earlier, scholars have observed that genetic science seems to reinscribe racialized categories, but less attention has been paid to the way genetics also transforms these categories and concepts. After all, genetics has been an important tool in contesting the biological concept of race, such that, if racialized concepts are reproduced in genomics today, they take a different form. Abu El-Haj (2012), for example, notes that anthropological genetics often uses noncoding genetic markers that are understood to be completely non-determining of phenotype, let alone culture, a fact that distances today's genetics from early twentieth-century concepts of race. While this is true for the kind of genetic analysis that is her focus and that looks at population history, more medically inclined genetics often looks precisely for links between diseases and ancestries. So the picture is uneven: racialized categories disappear and reappear, sometimes in quite familiar guises, but race is also refigured as genetic ancestry, which can be finely differentiated, multiple, and historically so distant as to be invisible and highly abstract (see conclusion, this volume). This mode of racialized thinking combines, in new ways, elements of reification and stabilization with processes of deconstruction and destabilization. Racialized thinking is an ongoing natural-cultural assemblage, which has always combined aspects of naturalization and determinism with aspects of culturalization and indeterminacy; genomics produces this ambivalence in a new way.

One aspect that the existing literature on race and genetics does not highlight, but which emerged in our data, is the role gender plays in intersections between genomics and ideas of race, ethnicity, and nation. The fact that discourses and practices of race, ethnicity, and nation are usually gendered and sexualized is well known and has produced an abundant literature.[8] But there is very little that examines how intersections of gender and race/nation operate in genomic research on human diversity (but see Hartigan 2013a; Nash 2012). Yet, as M'charek (2005a: 130) points out, the use of Y chromosome and mitochondrial DNA analysis—common in much of the research we studied—is "a technology for producing sexualized genetic lineages." This requires a brief explanation.

Some DNA tests (autosomal) estimate ancestral contributions by looking for markers in all an individual's DNA, inherited from his or her myriad ancestors. In contrast, analysis of mitochondrial DNA (mtDNA, inherited only from the mother) and Y chromosome DNA (inherited only by men and only from the father) focuses on genetic material that has been passed in a single continuous line of descent, either maternal or paternal. These tests look for specific mutations that occurred in an ancient individual and which have been passed on unaltered in unbroken lines of unilineal descent. Thus many Amerindian males share a mutation on the Y chromosome thought to have occurred in a male born in Asia about 15,000–20,000 years ago; among many indigenous populations in the Americas, around 50 percent of men have this mutation (see Genebase Tutorials 2013). Possession of this mutation means a male must be a descendant of the original ancestor and defines him as belonging to the Q haplogroup and as having indigenous ancestry—via this very specific unilineal connection. Belonging to the Q haplogroup indicates indigenous ancestry, but not belonging does not mean absence of indigenous ancestry, as this can be indicated by other genetic markers and by autosomal DNA tests. The point is that unilineal tracing of this kind is a powerful way of tracking prehistoric population migrations: these parts of the genome act well as "molecular clocks," because they establish a continuous lineage back into the distant past, with estimated mutation rates allowing deductions about the passage of time.

The maternal and paternal lineality of these parts of the genome means that they can effectively stand in for men and women (although men also have mtDNA) and for the sexual relationships between them, as of course DNA is transmitted through sexual reproduction. Their function as molecular clocks also enables narratives to be told about histories of sexual and gender relationships—although sometimes it is more a case of allowing existing historical narratives to be retold as genetic ones.

In our research, this was a consistent theme, as many labs found high levels of Amerindian markers in mtDNA and high levels of European markers in Y chromosome DNA. This was explained in terms of colonial patterns—already established by historians—of European men having sex with indigenous and, especially in Brazil, African women. Thus ideas about the birth of the nation through mestizaje were reiterated in a genetic idiom (Wade expands on this in the conclusion to this volume).

Mestizaje, Race, and Nation in Latin America

A major frame for understanding the material generated by our project is the ideologies and practices of nation formation through the gendered and racialized processes of mestizaje. "Mestizaje," or "mestiçagem" in Brazil, can be roughly translated as race mixture, and the word "mestizo" (*mestiço* in Portuguese) was typically applied in the sixteenth-century Iberian American colonies to the child of a European (usually male) and an indigenous person (usually female). The term is not purely biological in meaning and also carries connotations of cultural mixture. In colonial Spanish America and Brazil, complex ways of reckoning social status emerged, with Spaniards and Portuguese and other whites at the top of the hierarchy, slaves and black and indigenous people at the bottom, while the middle positions were occupied by a regionally and historically diverse set of intermediate categories, including *mestizo, mulato, castizo, morisco, pardo, libres de todos los colores* (free people of all colors), and many others. These were very indeterminate categories that served a heterogeneous set of social functions. Parentage was often an important factor influencing one's place in the hierarchy, but occupation, wealth, and reputation also counted and indeed could shape perceptions of parentage. Categories were highly fluid and contested (Bonil Gómez 2011; Forbes 1993; Garrido 2005; Gotkowitz 2011a; Jaramillo Uribe 1968; Katzew and Deans-Smith 2009; Martínez 2008; Schwaller 2011).[9]

With independence, formal discrimination against the intermediate categories was dismantled, slavery was abolished (although not until 1888 in Brazil), and elites attempted to create liberal societies along European lines. The special institutional status of indigenous people as colonial tributaries was abolished, but in many areas, the *indios* remained isolated and retained some specific rights before the law. During the nineteenth century, in varied ways, elites recognized that their new nations were populated largely by mestizos of diverse kinds and, in the context of new nation-building projects and European theories about race, this was often seen as problematic and an obstacle to progress (Appelbaum, Macpherson, and Rosemblatt 2003b; Schwarcz 1993; Skidmore 1974). However, ideas of what Stepan (1991) calls "constructive miscegenation" were already counterbalancing negative evaluations of mixedness and holding out the possibility that mestizaje could have positive outcomes. Already by 1861, the Colombian writer and politician José María Samper could write that "this marvellous work of the mixture of races . . . ought to produce a wholly democratic society, a race of republicans, representatives simultaneously of Europe, Africa and Colombia, and which gives the New World its particular character" (1861: 299). That Samper's travel writings also betrayed a powerful contempt for the mixed people he encountered in the flesh is not unusual: the abstract veneration of mixture was no bar to prejudice against dark-skinned mestizos, steeped in "barbarism" and languishing in the tropical lowlands (Wade 1999: 178).

By the early decades of the twentieth century, the active glorification of mestizaje as the basis of the national character, as distinctively Latin American and as at least potentially democratic was evident above all in Mexico and Brazil, where it became part of state discourse about the nation. In Mexico, the idea of mestizaje was promoted as the basis of the national character by intellectuals such as José Vasconcelos (1997 [1925]), politician, educator, and author of La raza cósmica, a book in which, contrary to dominant European thinking about the inferiority of mixed-race individuals, he argued for the potential superiority of mixed peoples (Basave Benítez 1992; Miller 2004; see also chapter 3, this volume). In Brazil, Gilberto Freyre, author of many books on Brazilian history and culture, also promoted the idea of a mixed tropical society with unique and valuable characteristics, particularly via his influential 1933 text Casa grande e senzala (Benzaquen de Araújo 1994; Freyre 1946 [1933]; Maio and Santos 1996, 2010; Miller 2004; Pallares-Burke 2005).

In Brazil especially, but elsewhere too, the idea of mixture was linked to the idea that race and racism were not important features of Latin American societies, claims made from the early decades of the twentieth century, often with a self-conscious look toward the segregated United States (Fry 2000; Marx 1998; Seigel 2009). The core idea was that in societies in which extensive race mixture had blurred clear boundaries of racial categorization, racial stratification and racial identity were not as important as they were in the United States, South Africa, or Europe. This was sometimes phrased as a claim to "racial democracy," a slogan often associated with Brazil (Guimarães 2007; Twine 1998), although similar sentiments can be found in Colombia. Politician and writer Luis López de Mesa (1970 [1934]: 7) wrote that Colombia was no longer "the old democracy of equal citizenship only for a conquistador minority, but a complete one, without distinctions of class or lineage [estirpe]." Meanwhile, in various countries, immigration policies sought to attract Europeans and limit the entry of black people (Graham 1990).

Black and indigenous people, not to mention the working classes, did not necessarily believe formulations of racial democracy and sometimes made their views known through various political and cultural mobilizations (Andrews 1991; Sanders 2004). From the middle of the twentieth century, they were joined by academic critiques that challenged the existence of a racial democracy, particularly in Brazil: evidence mounted that racial discrimination is a significant factor in Brazilian and other Latin American societies (Bastide and Fernandes 1955; Telles 2004; Wade 2010). Academic critiques also began to reread mestizaje as an elite ideology that marginalized or erased black and indigenous people, while aiming toward a whiter future (Gall 2004; Gómez

Izquierdo 2005; Rahier 2003; Stutzman 1981). More recent approaches have understood mestizaje as being subject to contradictory readings, such that it can be appropriated as a subaltern discourse as well, and may make room for certain constructions of blackness and indigenousness rather than simply erasing them (Appelbaum, Macpherson, and Rosemblatt 2003b; De la Cadena 2000; Gotkowitz 2011a; Mallon 1996; Wade 2005). There are multiple mestizajes, rather than a single ideology or process.

In much of Latin America since about 1990, there have been important shifts toward official multiculturalism, manifest in political and other legal reforms, which have given new recognition and rights to indigenous and Afrodescendant minorities (Sieder 2002; Van Cott 2000; Yashar 2005). This multiculturalism was a significant departure from the attention long paid to indigenous minorities by many nation-states, which was often from the perspective of a paternalist ideology of *indigenismo* (indigenism), which celebrated the indigenous past but also sought to assimilate contemporary indigenous groups (Knight 1990; Ramos 1998). Brazil and Colombia stand out in these reforms, especially in relation to Afrodescendant groups (French 2009; Htun 2004; Restrepo and Rojas 2004; Wade 2002a). The multicultural shift has generally been understood as a significant move away from official ideologies of "the mestizo nation," especially in countries such as Brazil, Colombia, and Mexico, where the ideology had been strongly developed over the late nineteenth and twentieth centuries. Some scholars have questioned whether the shift is quite so significant, especially as it seems to fit rather neatly with neoliberal agendas, raising questions about whether official multiculturalism really adds up to much in terms of changing racial and ethnic political and economic hierarchies (Escobar 2008; Hale 2005; Rahier 2011; Speed 2005). But there is no doubt that multiculturalism has altered the public landscape of politics, culture, and the national imaginary, with indigenous and Afrodescendant groups having greater visibility, nationally and internationally. This landscape is important as a frame for contemporary genomics, as this volume shows.

Multiple Mestizajes and Taxonomies

The multiplicity of mestizaje is also manifest in the different ways in which ideas about nationhood and mixture have developed in different areas of Latin America—a diversity that may transect the nation itself. For example, in Mexico, the key process of mixture is considered to have taken place between Europeans and Native Americans (now estimated to be about 10 percent of the population, widely spread across the national territory), resulting in the dominance of a dualistic taxonomy that divides indigenous and mestizo.[10] The indigenous category is divided by ethnic identities and labels, such as Maya or Zapotec. Although Africans were present in colonial New Spain in significant numbers, they were marginalized in representations of the republic (Aguirre Beltrán 1946). Recent shifts toward multiculturalism have focused on rights for indigenous groups, but there has been some attention to the "third [African] root" of the nation (Hoffmann 2006). Afrodescendant people tend to be referred to as *negros* or *morenos* (browns). Although most people would identify simply as Mexicans, it is taken for granted that this is more or less synonymous with mestizo—to the extent that Afrodescendants there have recently been labeled Afromestizos.

In Brazil, the role of blackness in ideologies of mixture is greater.[11] Cultural icons, such as samba and carnival, which are associated strongly with Afro-Brazilian roots, became national icons decades ago. African-influenced religious practices, such as candomblé, are also part of urban popular culture. Indigenous people are about 0.5 percent of the national total, but the idea and image of the *índio* is nevertheless strong and figures importantly in ideologies of mestiçagem, although the image of the *índio* is associated with Amazonia and thus seen as distant and exotic. From the 1930s, the image of Brazil as a mixed, tropical society became part of the official representation of the country. Still, it has been one of the few Latin American countries to include a "color" (or more recently a "color/race") question in its census (Morning 2008); partly as a result, public discourse referring to color is more common than in Mexico or Colombia. In the 2010 census, 48 percent of people identified as *branco* (white), 43 percent as *pardo* (brown), and 8 percent as *preto* (black), with the remainder as *amarelo* (yellow) and *indígena*. While pardo is the bureaucratic label for mixed people, the term "mestiço" may be used in everyday parlance. Ideas about mixture vary within Brazil: the far south, for example, is very "white," while the northeast is rather "black."

Brazilian social taxonomies are particularly heterogeneous. Classificatory systems based on race, color, and descent coexist, and the same category can be part of multiple systems (Fry 2000; R. V. Santos et al. 2009). There is a tension between bipolar principles of classification—white versus nonwhite—and systems with multiple intermediate categories. Telles (2004: 87) identifies three overlapping systems: the census categories, which are in popular use; everyday categories, which make extensive use of the term "moreno"; and state and black social movement practices, which tend to oppose white and black. The latter are related to multiculturalist reforms, which have extended rights for indigenous groups and implemented affirmative action programs in the form of racial quotas in university admissions for self-identified black students (which may include pretos and pardos).

In ideologies of Colombian nationality, blackness is more important than in Mexico, but less so than in Brazil.[12] There is a small indigenous population (3.4 percent), but, as in Brazil, it figures large in ideas about the nation and its history—and in multiculturalist reforms. Indigenous groups are spread across various parts of the country, not concentrated only in the Amazon region. The idea of internal difference is a powerful trope in images of the Colombian nation, which is often talked of as a "country of regions," each with a supposedly particular identity. Whiteness and mestizoness are often associated with each other and located stereotypically in the highlands, whereas blackness is stereotypically associated with the Pacific and Caribbean lowland coastal regions; indigenousness is seen as both highland (Andean) and lowland (Amazonian) and is stereotyped as rural. In contrast to Brazil, mestizos are more likely to be assimilated to whiteness than to blackness.

These multiple mestizajes and taxonomic practices are important. While all three countries share a great deal in terms of underlying ideologies of mixture as the foundational substance of the national body, there is no single discourse or process here. The way genomics operates with racial and national categories reflects this, although not in a simple methodologically nationalist way; there is no straightforward homology between nation and scientific practice.

The Concept of Race in Latin America

In ideologies and practices of mestizaje, the term and concept "race" have been ambiguous in these three countries. In the early decades of the twentieth century, the term was widely used by Latin American writers to refer to entities as varied as la raza argentina; la raza blanca, negra, and india; and la raza iberoamericana (Graham 1990). It was also used more generically, for example, in the book Los problemas de la raza en Colombia (Jiménez López et al. 1920). The range of reference of the term indicates its ill-defined character, but the concept behind such usages—which date back to the nineteenth century—was similar to the widely accepted understanding of race in contemporary racial science and eugenics, as a natural-cultural category. Faced with highly mixed populations, Latin American thinkers tended to avoid the more biologically determinist approach to race of some Anglo-Saxon theorists, which condemned mixed peoples to inferiority, and instead took a more eclectic approach, inflected by Lamarckian views, which left greater room for improvement through social hygiene (Restrepo 2007; Schwarcz 1993; Stepan 1991).

This approach fed into an early tendency to avoid the term "race"—even if racialized thinking was arguably still present. While scholars have noted a general postwar trend in Europe and the United States toward a "neoracism" or "cultural racism," in which an explicit discourse of biology recedes, to be replaced by a discourse about culture (Balibar 1991; Stolcke 1995), it is important to recognize that, in general, culture and biology—or culture and nature—are always intertwined in racial thinking, so it is misleading to think in terms of a simple temporal transition from one to the other (Gotkowitz 2011b; Wade 2002b). Still, there are changes in emphases, and in Latin America a shift toward a more explicit reference to culture occurred earlier than in Europe or the United States. In Mexico in the 1920s and 1930s, eugenicists often referred to mestizos without explicit reference to race or racial groups: the "unraced subject or the generic mestizo" was discussed (Stern 2009: 163); an explicit discourse of race was absent, even though it was evoked by the very term "mestizo," which fused ideas of racial and cultural mixture (Hartigan 2013b). A culturalist approach to race was evident in Peru from the 1920s, in references to the "soul" or "spirit" of a given group (De la Cadena 2000: 19, 140)—as was also the case in Colombia, alongside use of the word "raza" (Restrepo 2007: 53). In Brazil too, Gilberto Freyre, strongly influenced by Franz Boas, separated race from culture, emphasized the role of the environment in shaping people, and downplayed "purely genetic" (i.e., racial) effects (Freyre 1946 [1933]: 18; see also Pallares-Burke 2005; Benzaquen de Araújo 1994; Maio 1999). Although it was not until after World War II that the explicit terminology of race was "largely abandoned" (Appelbaum, Macpherson, and Rosemblatt 2003a: 8)—at least in public political discourse and even then with certain exceptions in Brazil—important tendencies in that direction date from some decades earlier. This was allied to the ideas, noted above, that race and racism were not significant social issues.

Challenges to ideologies of racial democracy were made from early on and recently have gained ground in countries such as Brazil and Colombia, where the state has admitted that racism is an issue. This has led to an increase in public debates about race and racial disparities. In Brazil the census question about color changed in 1991 to one asking about color or race, while in 2010 Congress passed a statute on racial equality (after ten years of debate). Debates about public health and educational policies have openly talked about race as a factor influencing health, educational, and social inequalities and have targeted the black population as in need of special attention (Fry 2005b; Guimarães 1997; Maio and Monteiro 2005; Maio and Santos 2010). In Colombia, the 2005 census, although it used the word "ethnicity" and did not mention race, effectively lumped together all "Afro-Colombians" and all "indígenas": the statistics it generated have been used to address issues of "race and rights" (Rodríguez Garavito, Alfonso Sierra, and Cavelier Adarve 2009).

Despite its greater public presence in recent years, the use of the concept of race to talk about identity and social differences can still be contentious (Restrepo 2012: 181). In Latin America, saying that one is doing research on something called race—whether in relation to social mobility or genomics—may produce denials from a range of people that race is an appropriate object of attention. In Mexico, differences are rarely discussed in terms of race, whether in popular or political discourse, while ethnicity is seen as a term applying mainly to indigenous groups and usually combining notions of linguistic and cultural difference with economic poverty. In Brazil, the heated debates about the appropriateness of race-based affirmative actions in health policy and university admission are phrased not only in terms of whether they contravene principles of meritocracy, but whether they derive from an imposition of foreign racial categories (such as U.S. ones) and also threaten to strengthen racial identities and differences seen as having traditionally played a minor role: "It is not just social policy that is at stake, but the country's understanding and portrayal of itself" (Htun 2004: 61).[13] On the other hand, in Colombia, some recent vox populi journalism showed people in the streets of Bogotá confidently asserting that different races existed, in terms of biological differences, and in many cases claiming to belong to one of them.[14] In short, the term and the concept are both present and absent, invoked and denied, at the same time. Although multiculturalism has given the term and the concept greater currency, above all in Brazil, race still provokes ambivalent reactions. This is an important context for understanding the way racialized categories circulate in genomic science practices in Latin America.

Gender and Mestizaje

Mestizaje is a sexualized and gendered practice and ideology: it refers to sexual relations and reproduction between men and women perceived as belonging to different races, colors, or ethnicities (however those words are understood in the Latin American contexts). As noted earlier, genomic research often finds evidence in today's populations that reflects early colonial matings between European men and indigenous or African women, but it is important to grasp some of the cultural meanings that attach to this encounter and its historical developments.[15]

Relations of power obviously structured early sexual relationships, whether these were consensual, constrained by circumstances, or forced (i.e., rape).

Although white men initially had relationships with indigenous and African slave women, as time went on they more likely had sex with the growing population of free black and the dark-skinned, plebeian mestizo women, although where slavery persisted, slave women also would have been present. There is no doubt that sex ratios were skewed among Europeans, early on, but European women gradually formed an important part of elite society. White men protected the (sexual) honor of wives, sisters, and daughters, and the legitimacy of their offspring, while also having relationships and children with lower-class dark-skinned women, deemed to be without honor—being mestizo was considered tantamount to being illegitimate in colonial Iberian America (Johnson and Lipsett-Rivera 1998; Stolcke 1994). Male honor was not besmirched by these relationships. This so-called dual marriage system was common across Latin America and the Caribbean and is found today in patterns of men having official households alongside unofficial families *en la calle* (in the street) (Smith 1997).

For darker-skinned plebeian women, having a relationship with a richer, whiter man could present possibilities of upward mobility. For darker-skinned plebeian men, such hypergamic relationships were much less likely, but still occurred, especially if such men became upwardly mobile first. Interracial sexual relationships were, and still are today, charged with meanings lent by these hierarchies of power and wealth (Caulfield 2000). In Mexico, writer Octavio Paz (1950) has pondered how the Mexican nation was founded on an original act of sexual violence or domination (*la chingada*), symbolized by the encounter between conquistador Hernán Cortés and the *india* La Malinche. The latter is seen both as traitor to her indigenous roots and heroic mother of the mestizo nation. In Brazil, the black man who "marries up" in color terms is generally assumed to be obeying motives of social climbing (Moutinho 2004).

Genomic research that highlights Amerindian markers in mtDNA and European markers in Y chromosome DNA and infers patterns of "asymmetrical mating" (Bortolini et al. 1999) or "sex-biased genetic blending" (Gonçalves et al. 2007) evokes this history and its cultural baggage, yet at the same time tends to telescope the history and gloss over the hierarchical meanings still in play today. This point is discussed in more detail in the conclusion.

Science Studies

As well as being a contribution to research on race, nation, and gender in Latin America, our book is an intervention in the field of science studies: our research involved lab-based ethnographies, interviews with scientists, analysis of their published work, and attention to how their work circulated beyond the institutional networks of genetic science, into the public sphere. We were interested in how ideas about race, ethnicity, nation, and gender entered into the work of genetics laboratories, which raised questions about the relationships between science and society in the production of knowledge.

For social studies of science, it is a commonplace that the social and scientific registers are not separate and that science is a social and cultural practice, albeit one of a particular kind (Fausto-Sterling 2000; Haraway 1989; Jasanoff 2004b; Latour 1993, 2005; Latour and Woolgar 1986). Latour (2005) argues that it is wrong to conceive of society or the social as a separate sphere or context that shapes or influences science: chains of associations link things (objects, people, words), which are not compartmentalized into such spheres, creating natural-social hybrids. Yet the science laboratory and scientific methodologies are specifically set up to exclude the influence of social and cultural assumptions that are perceived as potentially distorting the reliability of results; and scientists (and many nonscientists) generally have confidence that their methods can in principle lead to reliable and veracious findings. Specific findings may prove to be wrong, because of faulty assumptions, poor practice, or inadequate techniques, but science itself is assumed by scientists (and many nonscientists) to have the self-critical power to uncover such failures and achieve more truthful results. Scientists often recognize that the categories they use to organize their work, the assumptions they start with, may be derived from commonsensical understandings from outside the laboratory. But these are generally taken as a pragmatic starting point, a way into the nebulous world of cause and effect.[16] If different categories and assumptions can be shown to lead to a more useful result, scientists are likely to relinquish the old ones. Scientists may also explicitly hold to particular moral positions and political objectives that guide the overall direction of their work—for example, antiracism, social justice, equality of representation, or ecological sustainability (Bliss 2011; Bustamante, De La Vega, and Burchard 2011; Fullwiley 2008). But such positions are not held to shape the actual results that emerge from the practice of science in the laboratory. In that sense, nature and society are held apart by scientists.

Latour (1993) identifies this as a characteristic of "modernity"—that nature and society are separated by acts of "purification," while such separations are continuously dismantled by the actual practice of constructing associations and networks that assemble together things supposedly divided between nature and society. This does not mean that science simply produces false findings—the unveiling of processes of assembling is not a "debunking"

(1993: 43) or showing that things scientists call facts are merely social fabrications (Latour 2004). Rather it means that science should be seen as a process of the revealing of truths, through the labor of putting together complex assemblages of people, money, objects, facts, and words—things that are not separable into discrete categories of natural and social. Such assembling can be disputed among scientists, indicating that there may be more than one way to reveal truth and that some truths get more airtime than others; some assemblages may become stable, standardized, and taken for granted.

Social studies of science thus aim to show that the purified separations of nature and society and the teleologies of the progressive overcoming of bias and error are not so simple; that the detailed practice of science is continuously mixing nature and culture into natural-cultural combinations; that social and scientific things are combined, by the labor of diverse human actors and the affordances of diverse nonhuman things, into material-semiotic assemblages that cohere (or not) depending on how well and convincingly they speak to different audiences and do useful work for them (Latour 2005; Law 2008; Reardon 2008). This is not a simple matter of social categories—such as those of race, population, or region—penetrating the laboratory and shaping scientific practice, because the point is that these categories are already natural-cultural combinations, which have themselves been formed historically in a complex set of moving associations between scientists (natural historians, physical anthropologists, demographers, etc.) and politicians, engineers, administrators, writers, entrepreneurs, and all kinds of ordinary people. Social categories enter the laboratory, sure enough, but not as an extraneous social contaminant; they are already part and parcel of the assemblages that genetic scientists are putting together. The idiom of coproduction captures this insofar as it avoids giving primacy to either the social or the science and instead sees them as mutually constitutive, even as a boundary between them is continually reinscribed (Jasanoff 2004a: 21).

One way of showing this process of mixing and assemblage is to exploit the differences within science itself—something the comparative scope of our project is well suited to doing. When everything is stable and agreed upon, it becomes harder to see how things could have been different. The way natural-social categories enter into the practice of science can be revealed by demonstrating that a given scientific version of events could have been otherwise; and this can be achieved by showing that scientists disagree or, even more so, use the same or similar data to arrive at different conclusions. This internal variation is expected within—indeed is integral to—science: it can be seen as part of the progressive teleology by which truth, in time, triumphs over falsity. But social studies of science can also explore this variation to reveal the way natural categories are also natural-social assemblages, which may operate outside the laboratory too.

For example, looping back to the previous discussion on how race works in genomics, we can look again at the notion of population. There is debate in the study of human genetic diversity about whether to sample existing populations, which are usually defined by cultural criteria, such as language, identity, residence, history, and so on—or biogeographical ancestry, as we have seen—or whether instead to spread a homogeneous grid across a given geography and sample randomly within it, as might be done with, say, fruit flies (Nash 2012b; Pálsson 2007: ch. 7; Reardon 2008: 309; Serre and Pääbo 2004). Both methods can produce a picture of how human genetic diversity varies across space. In one sense, this is a technical issue about which method best represents the variation that might exist and any structuring of it. But social studies of science show that complex natural-cultural ideas about populations are deeply embedded in the practice of science (Fujimura and Rajagopalan 2011; M'charek 2005a; Reardon 2005). Populations are considered to have evolved and reproduced within a particular geographical niche. This is reflected in the fact that genetic research on diversity often samples only individuals whose grandparents were born in the population locality, thus effectively imposing a definition of what the population is. As well, specific populations, defined by social criteria, are generally accessed for scientific purposes via gatekeepers of various kinds, including local doctors, anthropologists, or community leaders, who can negotiate collective or individual agreements that samples may be taken (Reardon 2008). These practical but also profoundly ethical issues are based on fundamental notions of human social organization and cultural diversity, notions that do not apply to fruit flies. The grid method is not as easy to work with, as it does not assume particular social units that can be defined in terms of their internal coherence and their gatekeepers.

Starting with populations and producing genetic profiles of them tends to reproduce a model of more or less bounded populations (often the ones that you started with, thus creating an overlap between socially and genetically defined populations). Starting with a grid tends to produce gradients or clines of gradual variation and reduces the impression of located genetic populations; the absence of boundaries suggests the continuous movement and biological mixture of people between populations.

Thus the natural-cultural concept of population—a Latourian hybrid if ever there was one, circulating historically through government, administration, animal breeding, statistics, demography, genetics, and so on—shapes project design and sampling procedures, which in turn produce results that tend to reaffirm the concept of population, now in a genetic idiom. This might look unexceptional, even banal, except that a different point of departure—the grid—produces different results. One might be tempted to see the grid starting point as acultural: it is based on a seemingly purely scientific notion of random samples. But it is also a natural-cultural category, based on ideas about human homogeneity: it starts with a concept of humans as not segmented into groups, but as mobile and interchangeable, constantly moving and exchanging things, such as ideas, things, gametes.

In short, population is a stable device in human diversity studies; it is widely used and provides entry points and affordances that make it practical and powerful. It is something of an "immutable mobile" (Latour 1987) or a "boundary object" (Bowker and Star 1999: 296; Star and Griesemer 1989): it works between different scientific communities—social scientists, doctors, geneticists—and elicits common understandings, while also performing specific roles for each community and producing different objects—a located cultural grouping, a demographic set, a category of patients or potential patients, a genetically distinct group—which may, however, be translated into each other.

Diversity in scientific practice helps reveal the way natural-cultural categories such as population, race, ethnic group, nation, or region operate in the practice and productions of science and how science reproduces but also changes them. Studies of genetic science indicate how categories such as Mexican, Mexican American, African American, Puerto Rican, European, African, Dutch, and Turkish act as starting points for sampling human diversity and pursuing projects linked to the search for disease-causing genetic variants or forensic DNA matching (see, e.g., Fullwiley 2007a, 2008; M'charek 2005a; Montoya 2011). In the process, these categories, which are already natural-cultural or biosocial constructs, often become increasingly biologized and geneticized, a process that may then convert disparities of health to biological differences and pathologize certain ethnoracially defined populations (Montoya 2011: 185; see also Duster 2003a, 2006a; Fausto-Sterling 2004; Kahn 2005). But the same studies also reveal differences and controversies in this process. Some scientists may reject ethnic labeling of their samples but encounter resistance to this from scientific journal editors (Montoya 2011: 166–169); some geneticists may want to establish a link between African ancestry and propensity to suffer from asthma, but have to tussle with their own data when the data contest such a conclusion (Fullwiley 2008); forensic DNA testing can be challenged in court on the grounds that it is using ethnically biased samples to establish the probability of a link between a suspect and crime scene DNA

(M'charek 2000, 2005a: ch. 2). Scientific practice does not therefore simply re-produce existing categories in an automatic fashion; there is unevenness and contestation, which helps reveal how the categories are working in the labora-tories, acting and being acted upon.

Diversity is something that arose out of our focus on specific laboratories and particular genomic projects. This highlighted differences in agendas and approaches, even if many geneticists shared some goals (such as seeking out genetic variants underlying complex disorders). Comparative laboratory eth-nography is not very common in science studies (although see, e.g., M'charek 2005a; Fullwiley 2007a, 2008): our project suggests it can be a useful way into seeing how certain categories get assembled in multiple ways in localized sets of practices, which form part of extensive transnational networks.

The focus on diversity also sheds light on questions of power: some assem-blages and sets of categories achieve dominance over others. If a Colombian research team invents a set of ethnic or ancestral categories for classifying its samples, but ends up using simplified and standardized ones when submitting for publication in the United States, without, at first, really questioning why, this is because the standard categories have a taken-for-granted status (see chapter 5). If an overriding focus on the mestizo emerges out of current ge-nomic research in Colombia, compared to earlier interests in Afro-Colombian and Amerindian populations, this may well be related to a recentering of the image of mixedness in the context of multiculturalism's focus on black and indigenous minorities (see conclusion).

Comparative Approaches

The business of isolating units of comparison tends to gloss over interconnec-tions and interactions between the units that might locate them as part of a common network rather than contrasting instances (Gingrich and Fox 2002). Recent work in the history of science has emphasized the need for "'connected histories'—in contradistinction to comparative histories—which argues for connecting stories between empires and geographical regions" (Safier 2010: 138). Following critiques of "methodological nationalism" (Wimmer and Glick Schiller 2002), which challenge the taken-for-granted status of the na-tion as the unit of study and comparison, there has been growing attention to transnational flows, showing how what may appear as a self-contained cul-tural context has been formed through transnational exchanges (Matory 2006; Seigel 2009). In South America, for example, "creole elites forged nationalist accounts of the land and its historical remains." But "the political movements of nationalist science were in fact transnational. . . . Discourses of nationalist science crossed different states in South America" (Sivasundaram 2010: 156).

Our project makes use of a comparative framework by focusing on three nation-states, yet we try to avoid methodological nationalism by emphasizing the transnational flows of scientific knowledge that connect the three nations among themselves and with international genomic science, bringing scientists across Latin America, North America, Europe, and Asia into collaboration. We also show how some Latin American scientists evince postcolonial concerns about their position within hierarchies of international genomic science. And we highlight the internal diversity of science within each nation.

It is important to recall that we chose a small number of labs in each country. In all three countries, and especially in such a large one as Brazil, research on human genetic diversity was carried out in many labs and covered contemporary populations as well as ancient DNA, with questions arising from the fields of evolutionary genetics, demographic history, anthropological genetics, and medical genomics. The labs we worked with were chosen strategically to fit with our particular interest in issues of race, ethnicity, and nation. Our ethnographic focus meant more time was spent on one or two labs or projects than on others. Thus we necessarily have to be cautious in generalizing about the state of genomics in general in a given country, above all when we found different approaches even within the small number of labs that we observed. In that sense, we were alert to the problems of methodological nationalism, which might have drawn us to make broad national comparisons (see conclusion). On the other hand, it was tempting to see national contexts as providing an important frame for understanding specific labs and projects. This could not be ignored, of course, yet it would be easy to overstate the significance of, say, the debates about race-based affirmative action in Brazil as setting the context for genomic human diversity research in general. These debates were undoubtedly a factor in some research, but the transnational world of genomic science formed a rather different context, which also powerfully defined the agendas by which geneticists worked.

Structure of the Book

The book is divided into two parts. The first has three chapters that provide historical and contextual background to the study of human biological and, within that, genetic diversity in each country. In chapter 1, Santos, Kent, and Gaspar Neto give an analysis of the trajectory of studies in the fields of physical anthropology and population genetics on race, miscegenation, and human biological diversity in Brazil from the end of the nineteenth century to the present. The argument is that, although there have been profound theoretical and methodological transformations in the past 150 years from the point of view of scientific thought, the question of miscegenation has been a key driving point in the study of the biological diversity of the Brazilian population.

Chapter 2, by Restrepo, Schwartz-Marín, and Cárdenas, analyzes the historical background to studies of human diversity and racial difference in Colombia, highlighting the changing meanings attached to constructs such as "black" and the way the country has been seen as strongly regionally differentiated. Particular attention is paid to the Expedición Humana, an early (1988–1994) attempt to map, in genetic and cultural terms, the diversity of Colombia, along lines rather similar to nineteenth-century expeditions.

The Mexican team—López Beltrán, García Deister, and Rios Sandoval—focuses in chapter 3 on the Mexican National Institute of Genomic Medicine (INMEGEN), its emergence, and early initiatives to sample different populations in the country and create a map of Mexican genomic diversity. This is put into the context of concerns with the mestizo as a scientific object, which go back to the nineteenth century and traverse admixture studies done in the twentieth century by such as León de Garay and Rubén Lisker. The way INMEGEN presents its public face is then analyzed before tracing recent changes in the institute's priorities, after the publication of the milestone article "Analysis of Genomic Diversity in Mexican Mestizo Populations" (Silva-Zolezzi et al. 2009).

Part II contains three case studies drawn from the lab ethnographies and interviews with scientists carried out mainly by the project's three postdoctoral researchers, Kent, Olarte Sierra, and García Deister, in collaboration with three project research assistants, Gaspar Neto, Díaz del Castillo, and Rios Sandoval. Each study gives a detailed and focused insight into scientific practice in this field of genomic research. In chapter 4, Kent and Santos look at a research project, which started with the possibility of a genetic continuity between the extinct Charrua indigenous people and the contemporary Gaúcho population of the state of Rio Grande do Sul, a group seen as distinct both culturally and, it was hypothesized, genetically. The authors argue that, throughout the different phases of the research process, there was significant continuity in the central idea of a genetic association between the Charrua and the Gaúchos. Over time, this idea took on different incarnations and was affirmed with differing levels of certainty. This chapter reconstructs the road traveled by this central idea from initial hypothesis to final scientific conclusion, showing how the distinctiveness of the Gaúchos and the idea of genetic continuity appear through the data and their interpretation.

Olarte Sierra and Díaz del Castillo H.'s case study (chapter 5) concerns a Bogotá university genetics laboratory that undertook a project in the northeast of Colombia. The authors observed that, in this project, the population categories used to classify samples resulted from a highly dynamic negotiating process, in which the struggle to account for diversity and to reflect on the consequences of the scientists' own practice was pervasive. The scientists produced innovative classifications of populations, which differed from those commonly used in genetics research (typically the categories of mestizo, Amerindian, and Afro-Colombian). However, there came a point at which their discourses and practices were flattened and simplified. The categories not only became discrete and static but also returned to their typical form in order to interact with national and international peers. This process was then the subject of self-reflexive debate among the geneticists, who became more intent on reestablishing their own categories. This chapter highlights the often contradictory processes of innovation and transnational standardization.

In chapter 6, García Deister explores the laboratory life of the Mexican mestizo. Her chapter deals with the inner workings of the INMEGEN project to map the genomic diversity of Mexicans. She traces the voyage of the Mexican mestizo from the public domain, through the genetics labs, to bioinformatic databases. García Deister focuses on the transformations by which the Mexican mestizo, who started as a blood sample, is materially reconfigured into bytes (information in a repository or in a cloud). Neither the blood nor the DNA nor the data sets are proxies for the Mexican mestizo; rather, the mestizo is present in each and every configuration in a multiplicity of ways. The laboratory life of the Mexican mestizo extends beyond laboratory walls and into society. Once inhabiting the information cloud, the mestizo takes part in negotiations of belonging that may blur national boundaries and encourage broader ethnic affiliations, therefore debilitating the protective screen of genomic patrimony.

Part II ends with a synthetic chapter that draws together material from the three case studies and other primary and secondary data. Chapter 7 addresses the difficult question of how categories of human diversity operate in the practice of genomic science in the laboratory, from agenda setting through interpretation and publication of results. The chapter explores the way natural-social categories enter into the practice of science by demonstrating that a given scientific version of events could have been otherwise; and this can be achieved by showing that scientists disagree or, even more so, use the same or similar data to arrive at different conclusions. The focus here is quite specific, exploring the research process in the laboratory, from setting research agendas through sampling procedures to the interpretation of data.

The conclusion reflects on the big questions addressed by our research, assessing the extent to which categories of race, ethnicity, nation, region, and gender are challenged, reproduced, and transformed by the kind of genomic research we have explored. Wade addresses how genomic research relates to changing regimes of the imagination and governance of cultural diversity in these countries, including recent multiculturalist regimes. He assesses the implications of the imagined genetic communities that emerge from the genomic research for notions of citizenship and inclusion or exclusion. Finally he returns to the question of comparison to reflect on our own practice, drawing out the major findings from our research and relating them to ongoing debates in the field.

Notes

1   A decolonial view from the south does take racism as fundamental to colonialism and modernity (Restrepo and Rojas 2010), but the analysis of race need not take North American experiences as paradigmatic.

2   The etymology of the word is contested, but it probably derives from terms related to horse breeding (Contini 1959; Liberman 2009; López Beltrán 2004: 182).

3   Noncoding DNA (sometimes known as "junk DNA") is DNA that does not directly encode for proteins and thus was thought to have no direct influence on the organism's phenotype. A large proportion of an organism's total genome is made up of noncoding DNA. Recent research indicates that this supposedly junk DNA may have complex indirect but vital roles to play in gene expression (Skipper, Dhand, and Campbell 2012) .

4   In addition to the sources cited, see Franklin (2001), Goodman, Heath, and Lindee (2003), Pálsson (2007), Strathern (1992), and Wade (2007a).

5   Examples include the use of DNA testing to allow entry only to genetically authentic children of immigrants and genetic testing to identify children of disappeared parents in Argentina. See the website DNA and Immigration (http://www.immigene.eu/) and Penchaszadeh (2011).

6   Montoya, for example, argues that biology and genetics in particular are "reductive and deterministic" and that racial determinism fits right into these patterns (Montoya 2011: 28).

7   On geneticization, see the literature cited above and in note 4. See also Latour (1993, 2005).

8   For a discussion and references, see Nagel (2003), Wade (2009).

9 We are grateful to Joanne Rappaport for directing us to some of these references.

10 On Mexico, see Basave Benítez (1992), Bonfil Batalla (1996), De la Peña (2006), Hoffmann (2006), Hoffmann and Rodríguez (2007), Lomnitz-Adler (1992), and Mallon (1995).

11 On Brazil, see French (2009), Fry (2000, 2005a), Guimarães (1999), Maio and Santos (1996, 2010), Ramos (1998), and Telles (2004).

12 On Colombia, see Gros (1991), Leal León (2010), Mosquera and León Díaz (2010), Rappaport (2005), Restrepo and Rojas (2004), Uribe and Restrepo (1997), and Wade (1993, 2002a).

13 See Maio and Santos (2005), and the debates in the same issue of the journal *Horizontes Antropológicos*. See also Fry (2000), Guimarães (1999), Maio and Santos (2010), and Telles (2004).

14 A Colombian state TV channel, Canal Capital, aired these programs about race and racism in Colombia on 14 February 2011 (see Procesocensales 2011).

15 There is a large literature on this topic. For a guide, see Wade (2009). See also Martinez-Alier [Stolcke] (1989 [1974]).

16 As Montoya says of the U.S. geneticists he studied who use ethnically labeled population categories, these are starting points for exploring biological differences: "theirs is a pragmatic claim" (Montoya 2011: 162).

part I
history and context

chapter 1

From Degeneration to Meeting Point

Historical Views on Race, Mixture, and the Biological Diversity of the Brazilian Population

Ricardo Ventura Santos, Michael Kent,
and Verlan Valle Gaspar Neto

Let any one who doubts the evil of the mixture of races, and is inclined, from a mistaken philanthropy, to break down all barriers between them, come to Brazil. He cannot deny the deterioration consequent upon an amalgamation of races, more widespread here than in any other country in the world, and which is rapidly effacing the best qualities of the white man, the negro, and the Indian, leaving a mongrel nondescript type, deficient in physical and mental energy.

–Louis Agassiz and Elizabeth Cabot Cary Agassiz, A Journey in Brazil

None of the different types within the Brazilian population presents any stigma of anthropological degeneration. To the contrary, characteristics from all of them are the best that could be desired.

–Edgard Roquette-Pinto, "Nota sobre os typos anthropologicos
  do Brasil"

If . . . we contemplate the structure of our population, we can see that Brazil represents a true MEETING POINT. . . . Brazilians probably constitute the most genetically diverse group of human beings on our planet and are far beyond any attempt at synthesis. What we intend is simply to describe and celebrate diversity.

–Sérgio D. J. Pena, Homo brasilis

The three epigraph quotations, written during different periods over the last 150 years, reveal the long road traveled by physical anthropologists and geneticists in their interpretations of race, mixture, and biological diversity in Brazil: from an outright rejection of mixture as a source of degeneration to its effusive celebration. In consonance with dominant racial thinking of the time, Swiss naturalist and Harvard University professor Louis Agassiz drew on his travels to Brazil in 1865–1866 to offer an extremely negative view of the racial-biological composition of this country, as well as of its consequences for the future viability of its population. These views strongly influenced Brazilian intellectuals and politicians of the time. The second citation, from Brazilian physician and anthropologist Edgard Roquette-Pinto, represents the biological diversity resulting from the process of racial mixture in rather more positive terms. It stems from the 1920s, a period of marked nationalism in Brazil, during which physical anthropologists contributed significantly to the development of more positive interpretations of the Brazilian population that strongly opposed the earlier thesis of degeneration. Finally, the third and most recent citation, by molecular geneticist Sérgio Pena, celebrates the biological diversity of the Brazilian population as unique within a global context. Genetic arguments developed by Pena on the nonexistence of race and the inherently mixed character of all Brazilians have in the past decade played a prominent role in the heated public debates on affirmative action policies targeted at the mixed and black population in Brazil.

The main objective of this chapter is to analyze from a historical-anthropological perspective the trajectory of studies in the fields of physical anthropology and human population genetics on race, mixture, and human biological diversity in Brazil from the second half of the nineteenth century to the present. It does so by focusing on three key periods in the development of scientific thought about race and mixture in Brazil, as well as its articulation with debates about national identity: approximately 1870–1915, 1910–1930, and 2000–present. In particular, we analyze the work of three key scientists that are representative of each respective period: João Baptista de Lacerda, Edgard Roquette-Pinto (both affiliated with the National Museum in Rio de Janeiro), and Sérgio Pena (Federal University of Minas Gerais).[1] We consider these scholars representative both through their central position within academia and national debates, and through the alignment of their ideas with dominant thinking in their scientific fields at the time. Although this chapter reveals profound differences between these authors in terms of the methodology they employed, the content of their ideas, and the values attached to mixture, it also brings to the fore a number of continuities that run across the different historical periods analyzed here. These shared elements concern, in particular, the centrality of the question of racial mixture in defining the research agenda of physical anthropology and/or human population genetics in Brazil, as well as the often important role of knowledge generated by these scientific fields in the dynamics of the construction of a Brazilian national identity.

As argued throughout this book, the development of scientific thought on race and mixture does not stand on its own. It is intimately articulated with wider social processes, as well as public debates on race and national identity, both at a national level and within Latin America as a whole. Thus, an additional objective of this chapter is to place such scientific thought within this wider context. It does so, in particular, by exploring the significant correlations between the different periods identified here and the timeline of elite discourses on race in Latin America established by Appelbaum, Macpherson, and Rosemblatt (2003a).

Racial Degeneration and Whitening:
João Baptista de Lacerda, 1870–1915

The first period analyzed here covers a moment when many Latin American countries were young republics and had recently abolished slavery. Economies were dominated by the agricultural sector, and one of the main challenges was to find a substitute for slave labor. In Brazil, as elsewhere, intellectuals perceived their respective nations as racially heterogeneous, while interpreting racial differences as the natural basis of social hierarchy. In addition, far-reaching processes of racial mixture were interpreted negatively as a form of degeneration, damaging the fitness and productive potential of the nations' populations.[2] Such ideas were strongly influenced by European racialist theories that conceptualized Brazil as the prime example of the degeneration resulting from race mixture (Stepan 1991). In this context, policies aimed at the "whitening" of the population were increasingly understood as an antidote. Therefore, encouraging European immigration was seen as a possible solution, not only for labor purposes but also to stimulate what was imagined as the possibility of sociocultural and economic progress through Europeanization. Between the second half of the nineteenth century and the first decades of the twentieth century, approximately six million European immigrants found their way to Brazil, mostly to its southern region (Appelbaum, Macpherson, and Rosemblatt 2003a; Maio and Santos 2010; Schwarcz 1993; Skidmore 1974; Stepan 1991).

The career of physician and anthropologist João Baptista de Lacerda is closely connected to the development of research on physical anthropology at the National Museum of Rio de Janeiro from the 1870s onward.[3] At the time, Rio de Janeiro was the capital of Brazil and one of the main centers of research and intellectual reflection on race and the identity of the Brazilian population. Lacerda was initially coordinator of research in anthropology and became director of the National Museum in 1895 (until 1915).

Lacerda's early research focused on the indigenous races of Brazil. In the first volume of the *Archivos do Museu Nacional*, he published studies of indigenous craniology (Lacerda and Peixoto 1876) and dental characteristics (Lacerda 1876). Lacerda considered the school of French scientist Paul Broca (Lacerda and Peixoto 1876: 48) as the primary theoretical-methodological reference for anthropological research conducted at the National Museum. During this period, physical anthropology in Europe—and in France in particular—experienced intense growth, with the publication of a wide body of literature and the development of a plethora of instruments for morphological characterization of the human body (Blanckaert 1989; Gould 1996; Harvey 1983; Massin 1996; Stocking 1968). The research of Lacerda and colleagues was based on detailed descriptions of bone morphology and measurements with the aim of constructing "a history of the Brazilian fossil man" (Lacerda 1875). Central research questions included the number of indigenous races, their age, their specific anatomical characteristics, and whether they were indigenous New World populations.

As was common in the anthropological tradition of the second half of the nineteenth century, it was considered possible to infer the intellectual and moral attributes of individuals from their physical characteristics (see Gould 1996; Stocking 1968; Blanckaert 1989). Following this tradition, during a period of intense debates about the Brazilian workforce because of the imminent abolition of slavery (which occurred in 1888), Lacerda's analyses led him to deliver an unfavorable judgment on the position of the Indians in the racial hierarchy of Brazil, as well as on their potential to participate effectively in the construction of the nation. Cranial measurements provided, in his eyes, evidence of intrinsic biological conditions of inferiority: "the portion of the thinking organ amounted to miniature proportions" (Lacerda 1882c: 23). As Lacerda concluded on the basis of the study of three Xerente and two Botocudo men: "They are savage, do not have any type of art, and do not have an inclination toward progress and civilization. . . . The Indian is unquestionably inferior to the Negro with regard to physical labor. . . . We measured the muscular force of adult individuals with the dynamometer . . . and the instrument detected a force that is lower than that generally observed in white or black individuals" (Lacerda 1905: 100–101). The relatively unknown indigenous populations and questions about how to integrate them into the nation were a constant presence in Brazilian intellectual thought in the second half of the nineteenth century. Scientific research at the time offered highly pessimistic views of such populations, establishing a "contrast between the historic Indian, womb of nationality, Tupi par excellence, extinct by preference, and the contemporary Indian, member of 'savage hordes' that wandered through the uncultured backlands" (Monteiro

1996: 15). Research by physical anthropologists of the National Museum in the second half of the nineteenth century mixed racial analyses with evolutionary notions. By situating the Indians at the lowest levels of racial hierarchy, they echoed popular theses of racialist determinism promoted by influential European intellectuals such as Henry Buckle, Arthur de Gobineau, and Louis Agassiz (Skidmore 1974). During the same period, other Brazilian scholars, drawing on notions of race, evolutionary schemes, and criminal anthropology, conducted research on the black segment of the population, most notably the physician and anthropologist Raimundo Nina Rodrigues with his research on the population of the state of Bahia (Corrêa 1982). As such, they generated interpretative schemes that affirmed the inferiority of the indigenous and black races, well in line with contemporary scientific thought.

In the later stages of his scientific career, Lacerda shifted his attention toward the question of racial mixture and the mestizo population of Brazil. In 1911, he participated in the First Universal Races Congress in London (see Fletcher 2005) as Brazil's official representative. There, he presented his memoir Sur les métis au Brésil (On the Mestizos of Brazil; Lacerda 1911). This work became known for advocating a process of whitening, as Lacerda argued that Brazil was a racially viable nation because its population was on its way to becoming a white race.

According to Lacerda, in order to achieve the whitening of the Brazilian population, it was necessary to overcome a number of obstacles. The first one regarded the fate of the Indians and black people and, in particular, those whose vices "were inoculated into the white and mixed-races" (Lacerda 1911: 12). According to Lacerda, because of their inherent racial inferiority, these groups were destined to progressively disappear by the process of "ethnic reduction": as white racial traits were stronger than black or Indian ones, mixture would inevitably lead to the whitening of the population. The second obstacle concerned the enormous contingent of people of mixed race—the most difficult barrier, according to Lacerda. He described mixed-race Brazilians as physically inferior to blacks as well as morally unstable; intellectually, however, they were described as comparable to whites. According to Lacerda, by the process of "intellectual selection," the "generous slave owners" had encouraged those with a more intellectual propensity to participate in social life, generating a differentiated mixed-race population.

The primary point defended by Lacerda in his memoir is that Brazil would follow the path toward whitening because the mixed-race populations, in spite of not constituting a "stable race," tended to have children with white people through "sexual selection." This process was especially common in Brazil, where "procreation [did] not obey precise social rules, where mixed-race people [had] all the liberty to blend with whites" (Lacerda 1911: 8). In addition to this internal dynamic of racial transformation, Lacerda called attention to the role of immigration as a factor that accelerated the process of whitening by infusing populations with "European/Aryan blood."

Lacerda's memoir may be seen as an exercise in reconciliation between a Brazilian social reality of mixed races and scientific theories that disqualified mixed races. Whitening represented a path toward the redemption of the Brazilian population. During the congress in London, Lacerda predicted that within a century the black population would have entirely disappeared in Brazil. As Seyferth (1985: 96) points out, "the whitening thesis reflects the Republican elite's concern at the beginning of the century with regard to the problem of mixture and its implications within the larger context of Brazilian history" (see also Cunha 2002: 271–275; Schwarcz 1993; Skidmore 1974: 64–65).

The concept that whitening the nation's population was possible—which might have seemed unlikely to some European theorists, for whom racial mixture was by definition degenerative—is linked to the fact that the idea of whiteness in Brazil had—and still has in the present—different meanings from those circulating in Europe or the United States. Racial belonging in Brazil has traditionally been based much more on an individual's appearance than on descent. Although whiteness is often equated with Europeanness, light skin color is a more central criterion. Additional criteria such as culture, level of education, and social and financial standing all contribute to making racial affiliation relatively contextual; marrying a whiter partner has often been used as a means of social ascension for one's offspring in Brazil. A corollary of this approach to racial belonging is the assumption that there are no absolute borders between different races. At the time of Lacerda's writing, dominant thinking invariably constructed the white race as superior, both biologically and culturally and, while whiteness today continues to be contextually defined, it is still accorded a high cultural and aesthetic value.

Race, Mixture, and National Redemption: Edgard Roquette-Pinto, 1910–1930

The second period in the development of scientific thought on race, mixture, and the diversity of the Brazilian population corresponds roughly to the decades of 1910 and 1920. This was a period of the emergence of nationalist projects in a variety of Latin American countries, with objectives standing in opposition to European and North American theories that condemned racial mixture. Discourses that emphasized common features and harmony became increasingly central to debates on race and national identity. Intellectuals began to produce narratives that inverted the premises of the inferiority of mixed-race people. Although still influential in many circles, the ideas of racial hierarchy and degeneration that were common in previous decades lost much ground in Brazil and the rest of Latin America to emerging perspectives that conceptualized racial mixture in neutral or positive terms (Appelbaum, Macpherson, and Rosemblatt 2003a; Maio and Santos 2010; Santos 2012; Schwarcz 1993; Skidmore 1974; Stepan 1991).

In Brazil, one of the most influential scholars of this new approach was Edgard Roquette-Pinto, who succeeded Lacerda as the leading anthropologist at the National Museum of Rio de Janeiro. Like his predecessor, Roquette-Pinto started his career with research on indigenous populations and later turned to questions related to the mixed population of Brazil and its future viability. In 1917, he published *Rondonia (antropologia—etnográfica)*, a book resulting from his research on indigenous populations of central Brazil as part of the Rondon Commission expedition of 1912. In contrast to earlier anthropological research coordinated by Lacerda, which was more of the armchair type, Roquette-Pinto's study represented one of the first moves toward ethnographic fieldwork at the National Museum.

Rondonia differs significantly from earlier analyses and interpretations of indigenous people in Brazil. Although race does feature as a guiding principle in this work, Roquette-Pinto attributes to it a much less deterministic influence on other aspects of physical and social life. In his views, it was the "inferior, primitive, backwards" culture of the Indians rather than their biological characteristics that was the main obstacle to their incorporation into modern "civilization" and the Brazilian nation (see Lima, Santos, and Coimbra 2008): "We should not be concerned with making them citizens of Brazil. We all understand that the Indian is an Indian; Brazilians are Brazilians. The nation should protect them, and even support them . . . without imposing, so that spontaneous evolution is not disturbed" (Roquette-Pinto 1917: 200–201). These views were strongly influenced by French positivism, and in particular by the work of August Comte. A central premise of positivist thought was the understanding that human societies around the world were in different stages of evolution. Given the correct conditions, Indians would naturally progress toward more advanced levels on the evolutionary scale. This positivist vision, sustaining the notions of the "relative incapacity" of indigenous populations and their "tutorship" by the state, was profoundly influential in shaping Brazil's indigenous policy throughout the twentieth century (Diacon 2004; Skidmore 1974; Souza-Lima 1995).

From the 1920s onward, Roquette-Pinto became increasingly involved in the debates on national identity and racial mixture. His study of soldiers from different regions of Brazil, living in military bases in Rio de Janeiro, resulted in the publication of the "Nota sobre os typos anthropológicos do Brasil" (A Note on the Anthropological Types of Brazil; Roquette-Pinto 1929). Nationalist in its conception—it was planned as a contribution to the celebration of one hundred years of the declaration of Brazil's independence in 1922 (Castro-Faria 1952)—the study proposed evaluating "whether the . . . anthropological characteristics [of mixed-race peoples] show signs of anatomical or physiological degeneration" (Roquette-Pinto 1929: 123–124).

Roquette-Pinto answered this rhetorical question with an outright "no." He was particularly concerned with the ability to populate Brazil's extensive territory and to exploit its riches (1929: 119). In the introduction, he strongly rejected immigration policies and earlier anthropological approaches, which had historically prioritized "searching for, at the price of gold, white people, without discernment, without oversight" (1929: 119). According to Roquette-Pinto, such approaches had led to "the best elements of the nation" being abandoned to "indigence." His work represented an attempt to demonstrate that mixed-race Brazilians had the biological ability to populate and explore the riches of the nation. "The deficiency is not racial," according to him, but should rather be sought in a lack of adequate social and political organization at the national level (Roquette-Pinto 1929: 123).

In raising and answering the text's central question, Roquette-Pinto was not only arguing against earlier negative interpretations of the country's population by Brazilian academics, but also against more recent and highly influential theories on the limited biological and intellectual viability of mestizos developed in Europe and the United States. In the first quarter of the twentieth century, the dominant view in the global scientific field was that human races differed both mentally and physically from a hereditary point of view, and that mixture between individuals of distinct races was biologically harmful (Provine 1986; see also Barkan 1992; Gould 1996; Provine 1973). While showing important continuities with the anthropological tradition of the second half of the nineteenth century, this approach was based on a new modality of biological explanations: Mendelianism.

The rediscovery in 1900 of Mendel's laws—in particular the insight that hereditary traits were transmitted by genes—led to a widespread trend in human heredity research interpreting the most diverse physical and behavioral characteristics as resulting directly from genetic action (Bowler 1989: 270–281; Mayr 1982: 727–731). In the new Mendelian theories, the idea of degeneration and other negative evaluations of racial mixture were reactualized in genetic terms. Influential scientists such as the U.S. geneticist Charles Davenport and the Norwegian biologist Jon Alfred Mjoen warned of the many physical and psychological dangers resulting from the "disharmonious" mixing of individuals of different races (Davenport 1917, in Provine 1973: 791; Mjoen 1931: 3). According to Davenport, for example, "One often sees in mulattos an ambition and push combined with intellectual inadequacy which makes the unhappy hybrid dissatisfied with his lot and a nuisance to others" (Davenport 1917, cited in Provine 1973: 791). Such views strongly influenced some key Brazilian intellectuals in contemporary debates on the future of the nation and its population, such as Oliveira Viana and members of the eugenics movement such as Renato Kehl. Their views were generally pessimistic, affirming that Brazilians were doomed to backwardness due to their racially mixed nature (for a more extensive discussion of these authors, see in particular Maio and Santos 2010; Schwarcz 1993; Skidmore 1974; Stepan 1991).

Although Roquette-Pinto was strongly influenced by Mendelian schemes in his explanations of the wide racial diversity existing even within single families in Brazil (see, for example, 1929: 138–139; Santos 2012), he did not share the pessimism of Davenport and Mjoen in relation to mestizos' physical and mental viability. In "Nota sobre os tipos antropológicos do Brasil" he aimed to go beyond the denial of what he conceived as "bleak rhetorical fantasies" (1929: 147). Analyzing physical, physiological, and mental or psychological characteristics of "young men from all states, sons and grandsons of Brazilians, from 20 to 22 years of age, all healthy and subject to the same life conditions" (124), Roquette-Pinto concluded that "none of the different types of Brazilian populations has any degenerative anthropological stigma. On the contrary, all of their characteristics are the best that could be desired" (145). This applied as well to the highly stigmatized *mulatos*, given that "none of the characteristics studied . . . allowed them to be classified as types that have suffered a process of involution" (129). For Roquette-Pinto, the solution to Brazil's problems resided in creating the necessary conditions—in particular through education and health policies—to allow the "Brazilian types" to demonstrate their full potential: "anthropology proved that man, in Brazil, needed to be studied and not replaced" (147).[4]

The ideas of Roquette-Pinto were also strongly embedded within the nationalist movement that emerged in Brazil in the aftermath of World War I

(1914–1918) (Oliveira 1990; Skidmore 1974; Stepan 1991). According to Skidmore (1974), warfare in Europe had been a reminder that a country's strength stemmed from its capacity to mobilize resources: its people, land, and industry. For Brazilian intellectuals and politicians, this raised the question of Brazil's prospects for becoming an important nation on the world stage. During this period, the racial issue permeated the public debate over national identity. From the view of prevailing racial determinisms, Brazil's feasibility as a nation was limited; one of its foundations—its people—was perceived as having a weak (racial) constitution; it was lacking "biological coherency," since it was made up of "'raceless masses'—radical and terrifying heterogeneities instead of biological unities" (Stepan 1991: 105). The nationalism of the 1910s and 1920s was a search for ideological liberation from the shackles imposed by racist ideals (Skidmore 1974: 146). Oliveira (1990: 145) refers to a "militant" current in this nationalism, which "involved the search for a new identity and [whose] parameter was the refusal of the biological models underlying racist thinking."

An example of the materialization of "militant nationalism" was the so-called Pro-Sanitation League, a political-intellectual movement that, between 1916 and 1920, "proclaimed disease as the country's main problem and the greatest obstacle to civilization" (Lima and Hochman 1996: 23; Lima 2007). The intellectuals participating in the movement were opposed to racial and climatic determinism and considered rural endemic diseases the main obstacle to Brazil's project for redemption (Kropf, Azevedo, and Ferreira 2003; Lima 2007; Lima and Hochman 1996: 29–30; Santos 1998). Most of these intellectuals had participated in field expeditions to Brazil's rural areas, which they portrayed as isolated and abandoned by the country's elites. The intellectuals of the pro-sanitation movement perpetuated the image of Brazil as a sickly country, attributing to the sciences, and more specifically to medicine, an important role in the process of national reorganization (Castro-Santos 1985; Lima 2007; Lima and Britto 1996; Lima and Hochman 1996; Stepan 1991; Thielen et al. 1991).

Roquette-Pinto was closely involved with the national redemption project during the first decades of the twentieth century. By offering scientific evidence that countered until-then-dominant views of the racial inferiority of indigenous and, later, mixed populations, he contributed to the redefinition of the Brazilian population and its biological diversity in more positive terms. In addition, he reframed Brazil's key problems as lying in the social and environmental domain, rather than in the field of race or biology. As such, he increasingly came to advocate reformist policies that focused on health and education

(Roquette-Pinto 1927, 1933, 1942). The nationalist movement's anthropological interpretations of Brazilian populations drew in significant measure on Roquette-Pinto's work. As such, he contributed to paving the way for the open celebration of racial mixture and its result—the mestizo—that would grow in importance from the 1930s onward, in particular through the work of Gilberto Freyre (especially Freyre 1946 [1933]) and the nationalist policies of the Getulio Vargas administration. Such views later became consolidated as the ideology of "racial democracy," based on the argument that the intense and relatively consensual mixing of the three founding populations have blurred differences to such a degree that at present there are no clear-cut distinctions between them, only a racial continuum from the whitest to the blackest individual. This point has often been pushed further to argue that this absence of clear divisions allowed Brazil to become a racial democracy in which relations between people of different colors are relatively egalitarian and racism plays only a minor role. This perspective has been repeatedly echoed up to the present (Fry 2005a; Guimarães 1999, 2007; Munanga 1999; Sansone 2003).

Genetic Mixture and the Nonexistence of Race: Sérgio Pena, 2000–Present

A third and final period in which biological interpretations of the Brazilian population have played a significant role in debates on national identity is to the first decade of the twenty-first century. In April 2000, at the same time as the commemorations for the five hundredth anniversary of the discovery of Brazil by the Portuguese, geneticist Sérgio Pena and colleagues published "Retrato molecular do Brasil" (Molecular Portrait of Brazil).[5] As professor of the Federal University of Minas Gerais in Belo Horizonte, Pena published in the subsequent decade a number of further studies on the genetic profile of the Brazilian population, and in particular on the relative proportions of the African, Amerindian, and European contributions. His research placed a strong emphasis on the high levels of mixture at the genetic level, on what he considered biological evidence for the nonexistence of distinct races and on the absence of a significant correlation between genetic ancestry and physical appearance in Brazil.

Before discussing Pena's work in more detail, however, it is necessary to offer an overview of the profound conceptual transformations in scientific and social thought about race and human biological diversity in the intervening period. From World War II onward, race became gradually displaced from public discourse. In Brazil, as in most other Latin American countries, the unifying figure of the mestizo continued to occupy a central role in nation-building processes. Racial differences and a pervasive reality of racial discrimination became increasingly effaced from official discourse on the nation and its population. The idea that Brazil was a racial democracy had hegemonic status, despite the fact that academic studies, some done under the aegis of UNESCO and others carried out by foreign and Brazilian anthropologists and sociologists, were beginning to challenge the image of racial democracy (Maio 2001). The concept of race also increasingly became displaced from the political domain as a result of the centrality of class-based identifications, as well as its replacement with the less-charged concept of ethnicity.

In stark contrast to the earlier stages discussed in this chapter, during this period interpretations of the Brazilian population from a biological perspective did not play a significant role in national debates. This displacement was due to a combination of factors. First, interpretations from the social sciences were increasingly central in the construction of a national identity from the 1930s onward, in particular through the work of Gilberto Freyre. Second, eugenics were generally discredited in the wake of World War II. Finally, the UNESCO research and debate on race, which used Brazil as one of its principal case studies, resulted in moving biological interpretations even further away from debates about the identity of human populations, prioritizing explanations from the social sciences (Maio 2001).

From the 1980s onward, however, social movements that based their political strategies on differential racial and ethnic identities became increasingly influential in Latin America. In Brazil, the black movement and an increasing number of social scientists highlighted the continued existence of racial inequalities and challenged the notion of racial democracy. Their efforts resulted in the adoption of affirmative action policies in the public sector targeted at the black population, of which university quotas have become the most publicized. This reinsertion of notions of race into public policy has resulted in a countermovement by intellectual elites, political parties (mostly from the center Right, but also from the Far Left), and the mass media, challenging the appropriateness of race-based affirmative action (Appelbaum, Macpherson, and Rosemblatt 2003a; Fry 2005a; Guimarães 1999, 2007; Munanga 1999; Sansone 2003; see also the following discussion).

Within the scientific field, a similar gradual displacement of race took place in the aftermath of World War II. While at the beginning of the twentieth century the heuristic value of race was embraced, by the end of the century belief in the biological reality of the concept had largely been abandoned. Biological differences between different races first came to be interpreted in neutral terms, rather than serving as a support for hierarchical orderings. Gradually, the application of the concept of race to the human species became contested in itself, in particular with the rapid expansion of population genetics from the 1960s onward.[6] This field used analysis of blood types, enzymes, and proteins—the so-called classical genetic markers—in the analysis of human biological diversity, thereby displacing morphologically based analyses. In particular, Richard Lewontin's (1972) classic comparative study of the occurrence of blood types and other markers in populations from diverse parts of the world became influential. It revealed that the vast majority—approximately 90 percent—of the biological diversity of the human species was found within so-called racial groups, rather than between them. In addition, genetic research revealed, among other things, the common origin of the human species in Africa and the very limited number of genes that account for the phenotypical variety on which racial classifications are based, such as skin color and cranial morphology. As a result, by the 1990s the concept of biological race, while it still had some purchase as a way of talking about population differences, had lost most of its currency, and many biologists denied its validity completely (Reardon 2005; see also introduction, this volume). However, as discussed elsewhere in this book, the turn of the century brought a renewed focus on race in genetic research.

It was in this context of theoretical and methodological transformations in research on human biological diversity that human population genetics emerged in Brazil in the late 1950s and early 1960s (Souza et al. 2013). Since then, numerous studies with classical genetic markers have been conducted at an expanding number of universities to understand the formation and evolution of the composition of the Brazilian population from a genetic perspective (see Salzano and Freire-Maia 1967). This body of research came to be known as "racial mixture studies" and aimed in particular to establish the relative contribution of white/Caucasian, black/Negroid, and Indian populations to the gene pool of Brazilian populations. Key authors such as Francisco Salzano, Pedro Saldanha, and Newton Freire-Maia highlighted the importance of the genetic diversity of the Brazilian population for the understanding of admixture and interethnic relations on a global scale (Salzano and Freire-Maia 1967). While this approach continued to use the concept of race as a central analytical category, it also had an explicit antiracist message: biological differences between racial groups were neutral and could not in any way be used to sustain racist theories (Azevedo 1980; Freire-Maia 1983; Salzano and Freire-Maia 1967).

The 1990s saw a methodological shift toward analyses of molecular DNA.

With this shift came a rapid expansion of the scale of genetic ancestry research in Brazil. The central concerns of such research, however, largely remained the same: mixture, the biological diversity of the Brazilian population, and the relative contribution of its founding populations—now redefined in geographical rather than racial terms, as African, Amerindian, and European—to its contemporary gene pool. At present, several research centers focus particularly on these issues, including the Federal University of Rio Grande do Sul in Porto Alegre (see chapter 4), the Federal University of Pará in Belém, and the Federal University of Minas Gerais in Belo Horizonte. The research directed by Sérgio Pena at the third institution has had a particularly strong impact in the public sphere and in contemporary debates on Brazilian national identity.[7]

From the year 2000 up to the present, Pena and colleagues have published a series of studies with samples from a variety of regions of Brazil, analyzing mtDNA, Y chromosome, and autosomal DNA markers. The central objective has been to disentangle the history of the formation of the Brazilian population from a biological perspective. On the one hand, Pena's research reveals a discursive continuity with earlier mixture studies, in particular in its treatment of the composition of the Brazilian population as "unique and fascinating" because of its high degree of mixture. On the other hand, however, there is also an important element of rupture: Pena explicitly rejects the concept of race and for him the high levels of genetic diversity existing in Brazil are evidence of the nonexistence of race as a biological reality.

Through sequencing portions of mitochondrial and Y chromosomal DNA of approximately 250 self-identified white men from diverse regions of the country, Pena and his colleagues (2000) in the first instance presented a comparative panorama of the geographic distribution and patterns of the paternal and maternal ancestries reflected in these particular portions of the genome. First published in *Ciência Hoje*, the popular science magazine of the Brazilian Society for the Progress of Science, this study revealed that the paternal lineages found in the Y chromosome DNA of the sampled white men were almost uniquely European (98 percent). In contrast, the maternal lineages found in the mtDNA revealed a more complex situation, with 33 percent of Amerindian and 28 percent of African lineages.[8] Pena and colleagues placed particular emphasis on the "surprisingly high" Amerindian and African maternal contribution as evidence of the mixed nature of white Brazilians. They explained this pattern of "asymmetrical reproduction" (primarily European men with mostly African and Amerindian women) by referring to the history of colonization since the fifteenth century. "The first Portuguese immigrants did not bring their women, and historical records indicate that they quickly began the process of mixture with indigenous women. With the arrival of slaves, from the second half of the sixteenth century onwards, mixture extended to African women" (Pena et al. 2000: 25).

Sérgio Pena's work has maintained a specific message throughout the years: it has sought to demonstrate that, to put it colloquially, appearances are deceptive when confronted with genetic evidence. In particular, many of his studies reiterate the argument that in Brazil there exists only a weak correlation between people's physical appearance (their color or race, in Brazilian terms) and their genetic ancestry. A study of the autosomal DNA of approximately 170 samples from a rural community in Minas Gerais, published in 2003, made this argument explicitly. Revealing the significant overlap—and difficulty in distinguishing at the genomic level—between subcategories of individuals classified as whites, mixed, or blacks, the authors concluded that "in Brazil, at an individual level, colour, as determined by physical examination, is a poor predictor of African genomic ancestry, as estimated by molecular markers" (Parra et al. 2003: 177).

Through publications aimed at a wider audience, such as popular scientific books and columns, Pena has consistently sought to establish bridges between his genetic research and debates on race and Brazilian national identity, often citing authors such as Gilberto Freyre, Darcy Ribeiro, and Paulo Prado (Pena et al. 2000, 2009). Mass media interpretations of his work have repeatedly emphasized that his research would confirm interpretations of the Brazilian population by Gilberto Freyre, Darcy Ribeiro, and other authors associated with the racial democracy theory, by revealing both the high levels of admixture and the difficulty of differentiating between individuals in terms of race. In addition, in the "Retrato molecular," the authors presented genetic research as a potential antidote to racism: "if the many white Brazilians that have Amerindian and African mitochondrial DNA became aware of this, they would better value the exuberant genetic diversity of our population, and, who knows, they might construct a more just and harmonious society in the twentieth century" (Pena et al. 2000: 25). While in this earlier phase Pena emphasized that white Brazilians were not as European as they might think, in his subsequent work the focus shifted toward destabilizing associations between blackness and African genomic ancestry. His most recent work analyzes the autosomal DNA of samples from four of the five macroregions of Brazil with a set of forty insertion-deletion markers that Pena developed himself. Its central conclusion is that, while highly diverse, for the Brazilian population across all regions of Brazil European ancestry is predominant, even among "nonwhite"—that is, black and brown—individuals (Pena et al. 2011). Elsewhere, Pena attributed this to the "important population effect of the program of 'whitening' of Brazil promoted through the immigration of *circa* six million Europeans" (2009: 875).[9]

By this time, Pena had become closely involved in public debates on the role of race and racism in the production of social inequalities in Brazil. During the administrations of presidents Fernando Henrique Cardoso (1995–2002) and Luiz Inácio Lula da Silva (2003–2010), a number of affirmative action policies formulated in racial terms and targeted at black populations have been implemented, with particular emphasis on the areas of education, health, and the labor market. Some of these policies, such as the use of racial quotas for selecting students to enter universities, involve the need to identify specific beneficiaries in order to be implemented (black students, for example). In the area of health, there have been extensive discussions on the creation of specific sickle-cell anemia programs for the black population in Brazil. Such policies, and those related to access to universities, have triggered heated public debates, receiving fierce criticism from diverse segments of Brazilian society, including political parties, the mass media, and segments of the national intellectual elite. A central point of criticism argues that, in the case of Brazil, inequalities are more socioeconomic in nature than racial, in addition to affirming the difficulties of defining who is black in a country in which the boundaries of color or race are notoriously fluid (see, for example, Fry 2005a; Fry et al. 2007; Grin 2010; Magnoli 2009; Maio and Santos 2010; R. V. Santos et al. 2009; Steil 2006).

It is in this complex political scenario that Pena's work has contributed to the construction of arguments against affirmative action by presenting evidence suggesting the nonexistence of race at the genetic level, as well as the difficulty of clearly distinguishing white from black individuals. The potential for articulating genetic evidence with political arguments offered by Pena's ideas has been a central factor in reincorporating biological knowledge of the Brazilian population into debates on national identity, after this kind of knowledge had remained on the margins for decades. The mass media has been particularly important in popularizing Pena's research and in reincorporating genetics into the public debate. Imbued with the authority conferred by the "mystical" aura of DNA (Nelkin and Lindee 1996; Pálsson 2007), genetically based arguments have at times provoked fierce controversies between geneticists, social scientists, politicians, and spokespeople of the black community.

The new genetics (or genomics) has generated a techno-cultural revolution that has transformed technologies, institutions, practices, and ideologies in a progressively growing range of social domains (Goodman, Heath, and Lindee 2003; Haraway 1997; Lippman 1991; Maio and Santos 2010; Pálsson 2007; Rabinow 1992; Santos and Maio 2005; Wade 2002b). Knowledge and technologies based on the new genetics have not only redefined biological, cultural, and social loci in an individual's surroundings, but also reconfigured wide-reaching macrosocial, historical, and political relationships. The anthropologist Paul Brodwin (2002) has highlighted the growing intertwinement of the development of genetic technologies, sociopolitical relations, and the construction of differentiated identities in the contemporary world. Within the context of the growing value placed on genetics, historically recognized identity models can gain even more legitimacy or be negated by the results of DNA sequencing, and other proposals may emerge that were not socially recognized in the past. Regarding the identity politics of social movements, genetic evidence has been used both to sustain strategies of constructing essentialized ethnic or racial identities and—in contrast—as a means to undermine such strategies, in particular by social scientists. A notable example of the latter is the work of Paul Gilroy. In his book *Against Race*, Gilroy argues about the importance of taking recent genetic evidence into consideration in reflections on what he calls the "raciology crisis." Advocating for "deliberately and self-consciously renouncing" race as a method of categorizing and dividing humanity, Gilroy (2000: 17) emphasizes that the recent biotechnological revolution requires a change in understanding concepts such as race, species, personification, and human specificity. At the same time, however, Gilroy (2000: 7) also highlights the "utopist tone" of his argument and recognizes that his radical "anti-racial" position may compromise (or even betray) social movements and other groups whose legitimate claims rely on forms of solidarity and political action that are shaped by identities that were initially imposed on them by their oppressors.

In his efforts at establishing connections with debates in the social sciences, Pena has frequently made reference to Gilroy's book. He has emphatically denounced the notion of race, comparing it to earlier beliefs in witchcraft and taking on the mission to "dis-invent" race.[10] The titles of several of Pena's publications in social science journals are illustrative both of his ideas and of his efforts to contribute to social and political debates in contemporary Brazil, including "Reasons to Banish the Concept of Race from Brazilian Medicine" (Pena 2005), "Can Genetics Define Who Should Benefit from the University Quota System and Other Affirmative Action?" (Pena and Bortolini 2004)," and "The Biological Nonexistence versus the Social Existence of Human Races: Can Science Instruct the Social Ethos?" (Birchal and Pena 2011).

The public reception of Pena's research and arguments has been ambivalent from the start. In some circles, his "Retrato molecular" was celebrated for its potential to elucidate the biological history of the Brazilian population (see references in Santos and Maio 2004) and for offering "scientific proof of what Gilberto Freyre formulated in sociological terms" (Gaspari 2000), thereby reconfirming long-term interpretations of Brazilian society related to the idea of racial democracy. In contrast, a number of black movement activists interpreted Pena's research in negative terms for similar reasons. According to Athayde Motta, for example, the geneticists' studies represented a "scientifically supported simulacrum" of the myth of Brazilian racial democracy. Furthermore, he argued that the results would give rise to "almost infinite possibilities for manipulation," with the potential to be used for "a pro-racial democracy campaign . . . a political and ideological discourse whose primary function is to maintain the state of racial inequality in Brazil" (Athayde Motta, cited in Santos and Maio 2004: 351).

Since then, Sérgio Pena and his genetic arguments have increasingly moved center stage in the debate on racial quotas, as his views have been appropriated by politicians, social scientists, and the media as part of their construction of a position against affirmative action policies. Several social scientists have made use of the genetic evidence provided by Pena in order to argue against what they perceive as the increasing racialization of collective identities resulting from the implementation of affirmative action policies (see, in particular, Fry 2005a; Fry et al. 2007; Magnoli 2009; Maio and Santos 2010; R. V. Santos et al. 2009; Steil 2006).[11] The results of Pena's research have figured prominently in the antiquota manifesto launched in 2008 by a variety of academics, politicians, artists, and other public figures.[12] In 2010, Pena himself participated in public hearings of the Supreme Court on the constitutionality of racial quotas, during which he contributed to arguments against such policies by reaffirming his views on the nonexistence of race. As such, his political views and scientific research are embedded within a wider contemporary current within Brazilian society that emphasizes and values mixture, against a competing tendency toward the production of differentiated racial and ethnic identifications, as expressed for example in the discourses of the black movement and the design of affirmative action policies. In response, explicit rejections of Pena's genetic arguments have also increasingly featured in the argumentation of advocates of racial quotas. These have mostly focused on denying the relevance of genetics for racial inequities that had their origin in social processes, in a country in which racial discrimination is based on phenotypical appearance—mainly skin color—rather than on ancestry or genetics. In addition, a more elaborate critical engagement of Pena's ideas was featured in the pro-quota camp's reply to the antiquota manifesto (Nascimento et al. 2008).

Through these developments, genetics increasingly became a central element in public debates waged around national and racial identity, as well as around the constitution of the Brazilian population.

Concluding Observations: One Image, Multiple Interpretations

This chapter has revealed both important differences and significant continuities in the trajectory of research in physical anthropology and—later— human population genetics in Brazil over the last century and a half. In spite of profound theoretical and ideological shifts and methodological transformations, such research has continuously taken the issue of race, mixture, and the biological diversity of the Brazilian population as its central focus. Operating at the interface of science, race, and politics, it has—at least in the periods analyzed here—simultaneously drawn on existing social understandings about mixture and racial relations in Brazil, and offered scientific support for such understandings and the public policy proposals they inspired. In addition, the knowledge produced about the Brazilian population in museums of natural history and, most recently, in molecular biology laboratories has often been at the core of ongoing debates about Brazilian national identity.

Both the shifts and continuities in the trajectory of research on the biological diversity of the Brazilian population are well illustrated by the multiple interpretations projected upon the well-known painting *A redenção de Can* (The Redemption of Ham, 1895), by the Hispano-Brazilian painter Modesto Brocos y Gómez (figure 1.1; see also Maio and Santos 1996; Santos and Maio 2004). This painting represents four characters, with a mud house wall in the background, a common feature in poor regions of Brazil. Standing at the left, an old black woman is looking upward with her arms partially raised, as if she were thanking the heavens for having found grace. Sitting on the right and with his back partially turned is a man in his thirties. With a white complexion, his appearance resembles that of an immigrant from southern Europe. The center of the painting is occupied by a mother-son pair: the mother (phenotypically a mixed *mulata*) resembles a Renaissance Madonna with the baby Jesus (white skinned) in her lap. Brocos y Gómez painted the work seven years after the abolition of slavery in Brazil. *A redenção de Can* has usually been interpreted as an expression of the ideal of whitening: the old black woman thankful that her daughter, a light-skinned mulata (therefore, partially whitened), has married a white immigrant and borne a child with a white complexion (Seyferth 1985).

In 1911, Lacerda used Brocos y Gómez's painting to illustrate his arguments presented in *Sur les métis au Brésil* (Seyferth 1985; Skidmore 1974). Lacerda be-

Figure 1.1. *A redenção de Can* (The Redemption of Ham) by Modesto Brocos y Gómez, 1895 (National Fine Arts Museum, Rio de Janeiro).

lieved that Brazil was on the path to whitening: through mixture, it would be possible to solve Brazil's racial problem. According to him, in a hundred years black people in Brazil would no longer exist. From this perspective, the white child in the arms of his mulata mother in Brocos y Gómez's painting represents the incarnation of Lacerda's ideal. In April 2000, *A redenção de Can* resurfaced in Sérgio Pena and colleagues' "Retrato molecular." For the geneticists, the child in the painting and his family origins illustrated their arguments about the high levels of genetic mixture of the Brazilian population. Symbolically speaking, Pena's genetic interpretations suggest that the child, representing the white Brazilian population at the start of the twenty-first century, is far from being exclusively European. It may well be white in appearance, but genetically speaking—contradicting appearances—it is mixed.

These interpretations of the same image are as diametrically opposed as the first and third epigraphs featured at the start of this chapter. While Lacerda read this painting into the future in order to predict that Brazilians would all be white by 2010, Sérgio Pena's "Retrato molecular" read the same image backward in order to defy such prognoses and to reassert the important proportions of African and Amerindian ancestry in contemporary white Brazilians. If in the past the boy in Brocos y Gómez's painting symbolized the path toward redemption from "the evil of the mixture of races"—in the terms of Agassiz and Agassiz (1879)—in the view of present-day geneticists he became the embodiment of the biological heterogeneity of the Brazilian people, constituting what Sérgio Pena described as a "meeting point."

Notes

Parts of this chapter present material that have appeared in earlier publications, which develop arguments of a different order, in particular Santos (2012) and Santos and Maio (2004, 2005).

1   Due to considerations of space, the aim of this chapter is to offer a broad outline of the trajectory of scientific thought on race and mixture in Brazil, rather than a detailed treatment of the work of each scholar.

2   The notion of degeneration has a long and complex intellectual and political genealogy, which in considerable measure goes back to evolutionist and naturalist thought of the second half of the nineteenth century (Chamberlin and Gilman 1985; Pick 1989). It is a highly polysemical concept that has influenced the arts—in particular literature—as well as the fields of science and medicine, as is the case with psychiatry and physical anthropology. Closely associated to notions of decline, within the field of anthropology the concept of degeneration became central in theories on racial hierarchies and determinisms, as well as on the effects of racial mixture (see Stepan 1985).

3   The National Museum was created in 1818, but only started developing anthropological research from the 1870s onward with the appointment of Lacerda.

4   In his later work, Roquette-Pinto (1933) explicitly criticized Davenport's and Mjoen's interpretations of the effects of racial mixture.

5   The title of this article recalls Paulo Prado's (1928) Retrato do Brasil, a text that interrogated the national character in historical and cultural terms.

6   As Science and Technology Studies scholars have recently indicated, this transition was far from being an abrupt one, with the persistence of racial and typological approaches in the emerging field of human population genetics (see Reardon 2005).

7   In particular, Sidney Santos of the Federal University of Pará has also conducted systematic research on the genetic profile of Brazilian populations. After initially focusing on Amazonian populations in the 1980–1990s, he and his main collaborators have increasingly shifted their attention to the national scale. Just like Sérgio Pena, he has developed his own sets of genetic markers, for both forensic analysis and ancestry research, including a set of forty-eight autosomal insertion-deletion markers. As discussed briefly in chapter 7, there are significant differences between the research of Santos and Pena, in terms of both approach and results. The choice to focus on Pena in this chapter is based principally on his greater participation and impact on public debates in Brazil. The research trajectory of Santos is discussed in detail elsewhere.

8   The results of mtDNA and Y chromosome analysis were, in addition, published separately in two international genetics journals (Alves-Silva et al. 2000; Carvalho-Silva et al. 2001).

9   This perspective seems to confirm prognoses made in previous eras regarding the long-term process of whitening in Brazil, with the important difference, however, that these effects are conceptualized in terms of Europeanization and that they coexist with a persistent view of the Brazilian population as highly diverse. This emerging focus on the biological Europeanization of the Brazilian population in Pena's views is discussed in more detail elsewhere (Kent, Santos, and Wade 2012).

10 Pena borrowed the idea of "dis-inventing" from a song by a Brazilian musician (Pena 2008: 18).

11 Since 2005, one of the authors of this chapter—Ricardo Ventura Santos—has participated intensively in public debates on the implementation of affirmative action policies in Brazil (see Fry et al. 2007; Maio and Santos 2005, 2010).

12 See Adesões à Carta de Cento e Treze Cidadãos anti-racistas contra as leis raciais, Petition Online, http://www.petitiononline.com/antiraca/petition.html.

chapter 2

Nation and Difference in the Genetic Imagination of Colombia

Eduardo Restrepo, Ernesto Schwartz-Marín, and Roosbelinda Cárdenas

The project known as Expedición Humana, or Human Expedition, was a milestone in genetic research in Colombia. This project, which ran from the end of the 1980s through the first half of the 1990s, sought to explore the diversity of the Colombian population in terms of molecular genetics and aspects considered cultural. For this purpose, the researchers carried out multiple expeditions to (mainly indigenous and black) isolated communities in the peripheral areas of the country.

The Human Expedition reveals the way in which a significant number of genetic researchers imagined the relationship between difference and nation in Colombia at a particular moment. Although their ideas reproduced historically sedimented representations of difference, they also ushered in a new set of arguments and technologies that purportedly revealed a reality which had remained hidden until then. For these researchers, difference had become visible at the molecular level for the first time in Colombian history.

The Human Expedition was not the only noteworthy genetic research project in Colombia during the 1980s and 1990s. Other research projects in population genetics, which were mainly associated with Emilio Yunis Turbay—one of Colombia's pioneers in genetics—focused on the analysis of mestizo populations. Contrary to the Human Expedition's interest in isolated communities, they were primarily concerned with analyzing Colombians' genetic intermixture. This was in line with dominant notions of mestizaje, which conceived of national diversity in terms of spatialized variations of the mestizo's "triethnic composition" (black, indigenous, and European). In the 1980s, the ideology of mestizaje, which maintained that all Colombians are racially mixed—albeit in different proportions of the constituent racial groups—was dominant. Thus, much of the population genetics research that was carried out during this period took mestizaje as a given. This was the case even though Colombian national identity has not been as strongly associated with the notion of the mestizo as that of some other Latin American countries, such as Mexico. Even though few Colombians would readily use the term "mestizo" as the most immediate identifier to refer to themselves or the Colombian population at large, the idea that most Colombians are mixed circulated as common sense. This triethnic idea of diversity is in some, but not complete, contrast with post-1990 ideas of multicultural diversity, which envision the Colombian nation as a mosaic of more discrete population groups. Despite multicultural reforms, this commonsense understanding of Colombia as mixed remains strong today. Now, as then, when asked about the ethnic character of the nation, it is quite common to hear people make statements such as "Aquí en Colombia somos muy mezclados" (Here in Colombia we are very mixed).

We begin with a brief overview of the trajectories of genetic research on human populations in Colombia, and then provide an in-depth description of the Human Expedition program, examining its impact on the way in which the relationship between nation and difference is imagined in Colombia. We then contrast the Human Expedition to the other population genetics research projects undertaken by Yunis and his collaborators. Finally, we point to some substantive changes that are taking place in genetic research in Colombia as a result of a recent emphasis on forensics. In recent years, national genetic imaginaries have emerged from the pragmatic search for ways of identifying bodies in the context of an escalated armed conflict. Although the priorities of population genetics have shifted, forensic genetics also draws from and reconfigures past notions of molecular difference among the nation's populations.

Beginnings of Human Genetics in Colombia

According to its own protagonists, the history of genetics in Colombia emerged thanks to founding figures who organized the field around particular research lines and institutions.[1] The first of these founders is the medical doctor Emilio Yunis Turbay.[2] He describes his first years as a "self-taught" geneticist as an experience of running back and forth between labs in order to use the centrifuge and microscope, which were located on different floors (cf. Fog 2006). His foundational history is tied to the Universidad Nacional and to the first lab that did paternity tests in Colombia, the Instituto Colombiano de Bienestar Familiar (ICBF), at the end of the 1960s. Yunis was a pioneer in the field of clinical genetics in Colombia when the main goal of this burgeoning discipline was to investigate the possible genetic baggage of unknown or understudied diseases. Also, given his role in the institutionalization of DNA paternity tests within the ICBF, Yunis was critical to the beginnings of forensic genetics in Colombia.

At the same time, another foundational figure of Colombian genetics appeared on stage: Dr. Jaime Bernal Villegas. Bernal was the first Colombian to obtain a PhD in human genetics, and upon completing his doctoral studies at Newcastle University in the United Kingdom he returned to Colombia and became the driving force behind the Genetics Unit of the Pontificia Universidad Javeriana's (PUJ) Medical School in the early 1980s. Over the next decade, this unit became the Institute of Human Genetics (IGH-PUJ).

In 1979, Dr. Hugo Hoenigsberg and Dr. Helena Groot, who were part of this first generation of Colombian geneticists, founded the Human Genetics Lab at the Universidad de los Andes (LGH-Uniandes) with Maria Victoria Monsalve. This lab spent its first years doing population genetics research that utilized blood groups and enzymes "to establish genetic divergences between different ethnic groups in the country."[3]

Once institutionalized as a disciplinary field, human genetics labs produced a second generation of geneticists who began using new biological markers such as blood groups, antigens, and different protein types, in addition to mitochondrial and Y-chromosome DNA. This second generation, much larger than the first, has shown a strong interest in understanding the population dynamics of the Colombian nation, both for clinical diagnoses and for anthropological purposes. The interests of the second generation extend to new research fields such as the analysis of complex diseases (e.g., depression and diabetes), and forensics and criminalistics (although as noted above paternity tests had been done since the late 1960s). This second generation of geneticists transformed university spaces in significant ways. They created a discipline with a primarily clinical and academic profile and established human genetics as the legitimate expert field concerned with explaining the relationship between difference and nation in Colombia.

Although the institutionalization of human genetics was driven by researchers housed in various labs across the country, the Human Expedition program, which was spurred by the IGH-PUJ, left a deep imprint on the second generation of Colombian geneticists and the practice of genetics in Colombia more broadly. What began as a population genetics project carried out by a small group of experts turned into a large interdisciplinary research and service program, which was first known as the Human Expedition 1992 (1988–1992) and then as the Great Human Expedition (GHE; 1993–1994).

In Search of the Hidden Americas: The Human Expedition Program

The Human Expedition 1992: In Search of the Hidden Americas was an ambitious program that involved students and professionals from various fields, mainly based at the Universidad Javeriana. The main objective, as its subtitle suggests, was the revelation of a deep truth that had long remained concealed. In its own language, the program consisted of "expeditions" to marginal areas to "visit isolated communities" (many of which were indigenous) to undertake studies, accompanied by a medical and dental team that provided free care. Dozens of *expedicionarios* or expedition members arrived in these "remote" places in order to do research or participate as members of the "medical mission." The program aimed to learn more about and value the "diversity of the Colombian population." Although genetics was at the center of the program's objectives, this was not the only aspect considered.

According to its main promoter, Dr. Jaime Bernal, the idea for the Human Expedition emerged in 1987. In an article published in the national newspaper *El Tiempo* (5 May 1991) and in the internal project newsletter, *Boletín Expedición Humana 1992*, Bernal writes: "Four years ago [i.e., in 1987], sitting at a traffic light on my way to the hospital I wrote in my notebook: 'In search of the hidden Americas, Human Expedition 1992' and I knew then that I had found the way to expand the impact of genetics, the field in which I had been doing research and clinical work since I was a med student" (Bernal 1991b: 2).

The Human Expedition 1992 also emerged from the convergence of multiple factors and from the accumulation of several years of clinical and population genetics research undertaken at the Clinical Genetics Unit of the Universidad Javeriana. In a certain sense, the program was a natural outgrowth of an interdisciplinary research practice that had become common in the field of genetics in Colombia. As one of the geneticists at the IGH put it, "Beginning in the seventh year of fieldwork . . . we started to speak about the Human Expedition in order to provide ongoing research projects with a [common] conceptual framework. Already at this time, these projects went beyond the genetic realm and engaged with cultural aspects such as music, architecture, art, and sociology" (Gómez Gutiérrez 1998: 132).

Various external factors also help us understand why the program emerged and why it took the form that it did, including the growing visibility of molecular research on human populations. The Human Genome Project, which began in 1990, gave international genomic research an unprecedented push and fed the social and political imaginaries of genetics that circulated among nonexperts. After this, other large-scale projects such as the Human Genome Diversity Project and HapMap emerged.[4]

Given this international effervescence in human genetics research, it is not surprising that the Human Expedition 1992 program established relationships with parallel institutions in other parts of the world, creating a network of exchange and collaboration. One contemporary article stated: "The project [has] awakened an unprecedented level of interest from other research groups across the world, and we have signed academic collaboration agreements with these research centers" (PUJ 1992: 14). Amid the numerous relationships with international institutions, the relationship with the Academia Real de las Ciencias Exactas Físicas y Naturales de España is noteworthy. This collaboration is described as the Human Expedition's involvement in a project on the "biological genesis" of "our populations," meaning the populations of Mexico, Venezuela, Chile, Argentina, and Colombia.[5] This project, in which colleagues from these countries participated, sought to identify and examine "some biological markers and to involve isolated populations" (*Boletín Expedición Humana* 1992, July 1989, 4).

In Colombia, the growing worldwide interest in population genetics did not go unheeded. As human genomics became a cutting-edge field, research centers raced to position themselves at the top of Colombia's list of genetic institutions. Thus, by the early 1990s, the Universidad Javeriana's Human Expedition program was one amid a number of research projects on population genetics in the country. According to Catherine Ramos:

> In Colombia, there existed similar projects in other universities such as the Universidad de Antioquia, the Universidad del Valle and the Universidad Nacional, but their magnitude was similar to the first project Human Expedition 1992, and their dissemination reached scientific publications but did not go much farther. At the Universidad de Antioquia, Andrés Ruiz Linares, in the Medical School, was responsible for a project titled "Study of the genetic structure of the Colombian Amerindian population using classical and DNA markers." Ruiz Linares had worked with professor Cavalli-Sforza, who was the principal investigator of the Human Genome Diversity Project (HGDP), and together with Sergio Pena, a Brazilian scientist, he created a Latin American committee to promote the advance of the HGDP regionally. At the Genetics Institute of UNal, a number of projects related to the genetic structure of indigenous communities were carried out, but it appears that they did not have any relationship or dialogue with their equivalents at the Universidad Javeriana. (2004: 14)

Therefore, the main differences between the Human Expedition and similar investigations elsewhere appear to be its scale and the fact that its home institution got involved at a university-wide level. Although there did not seem to be much dialogue and exchange with other groups dedicated to population genetics research, the Human Expedition program established links with other entities such as the Instituto Colombiano de Cultura Hispana (Colombian Institute of Hispanic Culture): "We have signed an agreement between the Universidad Javeriana and the Colombian Institute of Hispanic Culture for the development of Human Geography in Colombia. This work intends to survey all of the available information regarding the diverse Colombian ethnic groups, their history, their culture, and their biological structure. We have initially designed six volumes of 500–600 pages each, following a single format that will collect the most important aspects of each human group" (*Boletín Expedición Humana 1992*, April 1990, 8). The interesting thing about this agreement is that it signals the central role that the Human Expedition program played in defining, at both the conceptual and editorial levels, a collection of texts whose primary objective was to describe Colombia's diverse "human groups."[6]

Another important factor that helps explain the emergence of the Human Expedition 1992 is the historical conjuncture of the Quincentenary: "The Universidad Javeriana announced the Human Expedition 1992, beginning on October 12 [1988]. With it we expect to carry out an interdisciplinary research process that will lead to true knowledge of what the Colombian population really is 500 years after the arrival of the Spaniards to the Americas. The Human Expedition is therefore one of the activities with which the Universidad Javeriana will commemorate the Fifth Centenary of the Encounter between Two Worlds" (Bernal 1989: 1). The explicit association between the Human Expedition and the commemoration of the Fifth Centenary is symbolized by the choice of 12 October as the official opening date for the program.

Many of the program's internal documents refer to the project's contribution to the construction of a genetic map of the Colombian population. By 1994 this objective was being presented in the following way: "The Human Expedition is an interdisciplinary research and service process that is centered on the Colombian population's genetic map, which strives to give a biological explanation of the current structure of our populations, understood not only as human settlements but as dynamic processes of interaction between man and his environment" (Bernal and Tamayo 1994: 33). This aim is then disaggregated into a detailed set of objectives:

- To provide scientific evidence that highlights the human and cultural diversity of our country.
- To search within our human groups for problems whose investigation can contribute to the production of universal knowledge.
- To examine the human history of our country with modern technologies that can generate or confirm historical hypotheses.

- To strive to generate awareness about our biological and cultural identity within the universal context. (Bernal and Tamayo 1994: 37)

Several aspects of the project's objectives are noteworthy. First, scientific intervention is seen as capable of highlighting the biological and cultural diversity of the nation. Scientists appear as key mediators who facilitate the "discovery" and valorization of diversity. Second, there is a taken-for-granted relationship between the particular "problems of our human groups" and the possibility of contributing to the production of "universal knowledge." This suggests that Human Expedition researchers believed in the existence of universal knowledge (which is common in certain epistemological frameworks and in science more broadly) and were convinced that certain problems of "our human groups" (the "our" in this sentence is clearly a nationalist figure) could be translated by the expert's labor into the language of the universal. A third idea is that modern technologies can be employed to answer historical questions and that they are a decisive source of truth capable of confirming or debunking hypotheses regarding the origins, kinship, migration routes, and characteristics of different populations. Genetic decoding appears as an unprecedented and indisputable archive of "our country's human history." Finally, using arguments that are founded in human genetics and purport to penetrate to the deepest plane of the individual and his or her molecular composition, Human Expedition researchers shored up claims for the singularity of Colombian cultural and biological identity. As we will see later on, these objectives repeatedly appealed to the idea of an "us," and contributed to changing notions of Colombianness by fostering an awareness of "the country's diversity." This emphasis on diversity was in line with an overall shift in ideas of the Colombian nation, which followed the 1991 Constitution and the concomitant multicultural turn that shaped the political and theoretical imaginary in the early nineties.

A few years prior to this, the IGH's research objectives had not been presented in these same terms. Some of the program's projects, such as the one titled Anthropo-genetic Studies of Isolated Colombian Populations, described its objectives as follows: "To undertake a joint anthropogenetic investigation in order to continue outlining the genetic structure of Colombian populations, which we have begun to sketch in our previous research projects" (Bernal 1991a: 1). Another example of this shift toward biological and cultural diversity is evident in the way in which Alberto Gómez, then director of the Clinical Genetics Unit lab, described the Human Expedition program: "The Human Expedition 1992 . . . seeks to identify the genetic foundation that defines the

Amerindian, black, and mestizo races that inhabit our territory, as well as the ethnography of the Colombian man" (Gómez 1992: 10).[7]

Three years after it began, the members of the Human Expedition 1992 program had taken approximately thirty fieldwork trips and visited thirty-four "indigenous and isolated communities" (PUJ 1992: 14), located in peripheral areas such as the Orinoco and Amazon river basins and the southwestern, Pacific, and Caribbean regions, largely in areas far from capital cities. Except for a few "black populations" in Chocó, San Andres, and Providence Islands, and a "campesino [peasant] community" in Saboya (Boyacá), all of the places visited were indigenous communities. In both qualitative and quantitative terms, the biological data collection, fieldwork, and associated analysis of the program were centered on indigenous groups.

The Great Human Expedition: The Genetics of Salvage and Visibility

In 1992, the Human Expedition program was expanded into the GHE. This project ran from 12 October 1992 to 13 July 1993, during which time nearly four hundred students and professors carried out sixty interrelated research projects. The first of the expedition's five phases began in Bogotá and headed southwest toward Tumaco, continuing through many of the most peripheral areas of the country. The GHE's expeditions visited "more than 50 indigenous, black, and isolated communities across the country. Data from 8,815 individuals belonging to the diverse ethnic groups that make up the Colombian population were collected. These were distributed as follows: 5,989 indígenas, 558 mestizos, 1675 negros and 593 colonos [settlers, colonists]. Among indigenous groups, 34 different ethnic groups were covered" (Mendoza, Zarante, and Valbuena 1997: 5).

The GHE included a medical and dental mission, as part of the interdisciplinary research program. The archives include a description of the medical supplies used while treating thousands of patients: "Given that providing communities with medical and dental care was another one of the Expedition's objectives, during the Expedition's 17 trips the following were distributed to patients: 150,400 antiamoebic capsules, 5,525 boxes of antibiotics, 28,000 analgesic tablets, 2100 antiparasitic treatments, and 25,500 vitamin tablets. And all this infrastructure enabled us to treat approximately 8000 patients in the most remote places of our country" (Zarante 2013).

As was the case with the Human Expedition 1992, the GHE also emerged in the conjuncture of the Fifth Centenary. This time, however, the breadth and reach of the GHE were presented as a herculean effort made by an interdisciplinary academic community in order to make visible the country's isolated communities, thereby making Colombians aware of their country's "multi-ethnic wealth." It was also envisioned as a means to build bridges between these isolated communities and other Colombians in an effort to find solutions for the former's urgent needs:

> The Fifth Centenary of the Encounter of Two Worlds was key in taking an important step forward in the Human Expedition in order to better make sense of all the knowledge that has been acquired over these years, turning Colombia's attention to the situation that our isolated communities live in, and looking for sources of solutions to some of their most important needs. For this, we have planned a Great Human Expedition that will cover all the previously visited territories in order to continue our research process, to enable our isolated communities to find interlocutors that can aid them in their self-empowerment process, and to produce a graphic archive that will give other Colombians a clear idea about their multi-ethnicity. (Boletín Expedición Humana 1992, 1991, no. 13, 7)

Following a style reminiscent of the two paradigmatic scientific projects of what is now Colombia, the Botanical Expedition of the eighteenth century and the Chorographic Commission of the nineteenth century, the GHE hired an artist who was charged with producing a register of the faces, places, objects, and situations witnessed in the various expeditions. Photography and audiovisual media were thought not to possess the artistic and aesthetic qualities necessary to adequately portray the achievements of a scientific expedition with historic pretensions.

The interdisciplinary nature, the commitment to service, and the significant participation of students as expedicionarios are examples of the university's commitment to the program, which is in part explained by the successful fund-raising efforts of the director, Jaime Bernal. Another characteristic of the program is the fact that its participants produced numerous representations of it (such as the repeated statement that the GHE was heir to and on the same plane as the Botanical Expedition and Chorographic Commission; see Gómez Gutiérrez 1998). This suggests a desire to be hypervisible, made manifest through numerous articles published in national print media, countless publications written for a wide array of publics (that range from an article in the most reputable international academic journal to books written in a language easily accessible to nonexperts), and the production of many other materials that were not necessarily linked to genetic research.

The notion of an expedition and the use of the term "expedicionarios" reveal the particular positionality of the program's designers as urban, middle-class, and highly educated people; a position from which the program's destinations are imagined as remote geographies and habitats of forgotten peoples ("isolated communities").[8] Many of the published reports are written in an adventurous tone that reveals the expedicionarios' sense of adventure when confronting the unknown: "We have made many trips since then; we have walked for days on end in Nariño, traveled to the most remote places of La Guajira in a four by four [vehicle], spent hours on a boat on the Atrato, Vaupés, and Caquetá Rivers; we have made long journeys on horseback and airports have become part of our daily routine" (Bernal 1991b: 2).

In numerous publications, this sense of adventure is evident in first-person accounts that portray a group of city people, the expedicionarios, penetrating some remote area in order to discover both with their own eyes and those of "all Colombians" geographies, natures, and "isolated communities" associated with the "Other Colombia," the one inhabited by the still "undiscovered Colombian."[9] These accounts narrate "anecdotes that contrast with the expedicionarios' urban perspective" and reveal how the expedicionarios discovered "places and people that inhabit a world that is entirely distant from the one in which they have routinely lived" (Ramos 2004: 8).

The results of the Human Expedition program are numerous. One of the most salient outcomes was the creation of a biological bank of human tissue (which in some articles is called an Amerindian biological bank, or American biological bank). This bank was created in the early 1990s in order to conserve the biological material that had been collected over the years from the work performed in the Clinical Genetics Unit and the Human Expedition, but it also provided a storage service to clinicians and researchers who wanted to deposit samples there (Gómez 1992: 11):

> At the Universidad Javeriana we have recently created the Amerindian Biological Bank that seeks to collect and store all of the biological samples that have been gathered during the process of Human Expedition 1992. These samples are of special scientific importance given how difficult it is to obtain them and how rare some of the found genetic disorders are. For this reason, they will be made available to anyone who has a research interest that has not been covered by the expedition's work. The Bank has a collection of nearly 2000 plasma and hemolized samples, and shortly we will make our sample catalogue public, along with a description of the biological characteristics that we have studied. A very interesting aspect of this Bank will be the immortalization of studied individuals' lymphocytes, which will enable

us to use live cells even years after the Human Expedition 1992 project concludes. (*Boletín Expedición Humana 1992*, November 1990, 1)

A year later, Alberto Gómez wrote the following: "One of the projects currently under way is what we have called the American Biological Bank. This project sets out to provide a safe storage place for the representative organic samples of the three races that make up this program" (1991: 8). Just a few years later, the existence of the bank, as well as the collection practices associated with it, would become the object of a fierce debate that was initiated by the indigenous movement and directly involved the Human Expedition program.[10]

The myriad research projects that emerged out of the Human Expedition program are one of its most notable results. Although some of these were undergraduate and master's theses, most of which remain unpublished, many of them (including some of the theses) have been published as books and articles. In his book *Al cabo de las velas: expediciones científicas en Colombia S. XVIII, XIX y XX*, Alberto Gómez Gutiérrez (1998) presents an annex with a detailed list of all the research project results linked to the Human Expedition (both published and unpublished). These studies range from (clinical and population) genetic research to the medical and biological sciences. There is also a much smaller number of research projects in other disciplinary fields.

Population genetics is one of the most productive research lines of the Human Expedition program and perhaps where its most ambitious expectations were placed. A national newspaper, for example, wrote, "From the findings [of the Human Expedition] the country's genetic map and human geography will be outlined, enabling us to develop a clear idea of *that other nation that is also Colombia*" (*El Tiempo*, 1 February 1993, emphasis added). Although we can't really say that the program outlined a genetic map of the country, some of the elements necessary for its creation were produced. Some of the publications that stemmed from the program present aspects of population genetics among indigenous and black populations (the latter were done in Chocó and Providence Island). In some instances, these cases are contrasted to one another and in others they are contrasted with "mestizos," "colonos," and "Caucasoid" groups (the categories of classification that were used are discussed in detail below).

For academics and activists interested in the study of black people in Colombia, the publication of the journal *América Negra* was perhaps the most visible result of the Human Expedition program. In August 1990 the journal's first issue—which was initially titled *América Negra y Oculta* (Black and Hidden Americas)—was publicly announced. A year later, on 4 July 1991, the first issue of *América Negra*, edited by Nina S. de Friedemann, Jaime Arocha, and Jaime Bernal, was officially presented. The Universidad Javeriana and the ICFES (Colombian Institute for the Evaluation of Education) publicly announced their commitment to financially support the journal. A total of fifteen issues were published between then and December 1998.[11] The journal was an international publication that accepted "manuscripts from any of the disciplines dedicated to the description of human populations, and that place particular emphasis on black and indigenous communities in the Americas, and on their relations with populations in other parts of the world" (*Boletín Expedición Humana* 1992, August 1990, 4). Over the years, a large number of the findings from the research projects associated with the Human Expedition program appeared in the pages of *América Negra*.

Other, less visible publications included the series *Terrenos de la Gran Expedición Humana* (Terrains of the Great Human Expedition), which printed a dozen books with results from investigations in the Human Expedition 1992 and GHE programs. Another publication was the series *Artes y crónicas de la Gran Expedición Humana* (Arts and Chronicles of the Great Human Expedition), which produced four issues, a CD titled *Itinerario musical por Colombia* (Musical Itinerary through Colombia), and the book *Diseño indígena* (Indigenous Design). Also, the Human Expedition program participated in the recently created Latin American Association of Biological Anthropology, the official publication of which is the journal *Antropología Biológica*. By 1992, Jaime Bernal was the association's secretary and the Universidad Javeriana's Press published the first issues of *Antropología Biológica* with Jaime Bernal as the journal's editor in chief.

In a letter addressed to Luis Guillermo Vasco, a professor at the Universidad Nacional, in response to a debate that had been unleashed regarding the genetic work of the Human Expedition program, Bernal presents an overview of the program's achievements as well as a description of the general framework within which it operated:

> I think that our research has enabled us to make a thorough diagnosis of the precarious nutritional, educational, health, dental, etc. situation of many of the country's most isolated communities (which we have disseminated at all levels and which is essential information to put in practice actions that adequately respond to these people's needs); to reveal, together with the studies performed by other researchers such as Dr. Emilio Yunis, our people's past, to cultivate a curiosity about the importance of our ethnic wealth (through

expositions of our artists' works, publications, student presentations, etc.);
to generate and support the furthering of knowledge production regarding
Afro-Colombian populations in the national and universal contexts (this
last intervention was done through the publication of *América Negra* and our
participation in multiple forums of various nature); and finally, to show
the marvelous biological and cultural diversity of our country, which I have
personally presented, as well as many of the rest of the project's participants,
before numerous international and Colombian audiences, and in which I
usually underscore the need to know ourselves if we want to live in reasonable harmony, because we cannot love that which we do not know, and we
cannot tolerate what we don't love. (Bernal 1996: 6)

This summarizes the way in which the creators of the Human Expedition as
well as its many associated expedicionarios saw the work they were doing. In
their view, this was not a narrow genetic science endeavor, but one that sought
to reveal the true history of the nation as well as the cultural and biological
wealth that was contained in its diverse human populations.

Nation and Difference

Nina S. de Friedemann, a renowned Colombian anthropologist who had been
doing research among black populations in different parts of the country since
the 1960s, joined the Human Expedition's team in the early 1990s. For Friedemann the Human Expedition was an opportunity to continue overcoming
what she called the "invisibility" of "black groups" in Colombia (Friedemann
1984):[12] "Within the university realm, black groups cry out for the creation of
specific educational and research programs comparable to those of the Indian
ethnic group, which have never been heeded by the same institutions that
teach anthropology as a science that seeks to explain man. . . . Fortunately,
this demand, made by black groups in Colombia, was heard by the Universidad
Javeriana's Genetic Unit, within the context of a research program that honors
its title, *In Search of the Hidden Americas*" (Friedemann 1990: 1, emphasis in original). In addition to the contrast between "black groups" and "the Indian ethnic
group," which appears to reproduce a clear racial distinction (black/Indian),
this quotation demonstrates that Friedemann's enthusiastic participation in
the Human Expedition must be understood within the context of her disputes
with the anthropological establishment and her efforts to position the study
of black groups within it, making visible their historical and cultural presence

and their contributions to the nation: "In all, this project's most important accomplishment within the trajectory of the Human Expedition 1992 . . . is the incorporation of black groups as suitable subjects of study of *The Hidden Americas*. The marginalization that black groups have suffered within the university and the realm of research will begin to decrease" (Friedemann 1990: 1).

Friedemann's concrete expectations of the Human Expedition are clearly stated in her presentation of a research project in which she participated directly, titled Ethnomedical and Genetic Profiles on the Pacific Coast:

> This interdisciplinary anthropological-genetic and pharmacology project constitutes an effort to understand the ethnomedical perspective of black groups on the coast, in the field of genetic pathology. Following this, we consider it viable to undertake an emic/etic interpretation of bioscientific concepts and methods. Of course, when the project was designed we kept in mind the possibility of examining some of the reasons that explain the Indian-black demographic asymmetry on the [Pacific] coast. Living in a strange environment and under the yoke of several centuries of slavery, blacks survived to such an extent that they changed the face of the coast from aboriginal to black. . . . The knowledge of the genetic structure of these groups could offer some clues to answer this question. Similarly, it offers the possibility of tracing some of this population's origins ranging from the African diaspora to the results of internal migration processes across the country and regional agglutinations. (Friedemann 1990: 1)

Friedemann's hopes of using genetic studies to trace the specific origins of African-descended populations as well as to identify kinship lines and genetic distances were founded on claims of the geneticists of the Human Expedition. For example, one of her colleagues in the project in Chocó, the molecular geneticist Ignacio Briceño, wrote:

> According to Germán de Granada's [sic] linguistic analyses, the origin of the inhabitants of the Pacific coast is Fantiashanti. Edward Bendix and Jay Edwards note that the San Andrés and Providence archipelagos have the same influence, and Carlos Patiño Roseli [sic] points out that the creole language of Palenque de San Basilio has elements of the Congo and Angola languages. Using studies of genetic markers that are found in HLA, the Human Expedition will contribute objective biological evidence that can help elucidate the genetic composition of these groups, and with it to highlight the biological and cultural identity of the Colombian people. To this end, we are doing

investigations in Chocó, where the samples taken will be compared with studies of Providence currently underway, and with those that are done in the future. (Briceño 1990: 4)

Jaime Bernal also highlighted genetics as a source of information for tracing the historical processes of settlement of the country's populations. For him, genetics allowed historians and linguists to paint the picture of "our country's" history and prehistory with a finer, though complementary, brushstroke than the one they had used until then. This is evident in the following excerpt, where Bernal explains the dazzling terrain of science to neophytes:

> Thanks to new technologies, biology allows us to bring our country's history to life in order to understand it and make it our own. Now, in order to do all this we need to study the genetic structure of each of these groups from different angles. Ranging from the simplest to the most complex, we want to typify them in order to know their blood groups, the variations in their serum and red blood cells, the different forms in which the HLA antigens are present, and finally, their variations in DNA sequence, both in the nucleus and in the mitochondria. All of these data enable the production of mathematical models of the relationship between different human groups in order to construct dendrograms or phylogenic trees, which, interpreted in the context of known cultural or linguistic facts, can yield that coherent vision of our country's prehistory that we search for. (Bernal 2000: 14)

For several of the authors associated with the Human Expedition, the study of human (biological and cultural) diversity was urgent: in their view, this diversity was at risk of disappearing in the face of technological transformations and the accelerated mestizaje (or race mixture) that confronted these isolated communities. One of the arguments for the creation of a human biological bank was stated in the following terms: "The conservation of all of this biological patrimony is truly urgent, given that the different ethnic groups run the risk of being diluted amid the progressive mestizaje of these cultures" (PUJ 1992: 16). In a research proposal presented to the state science funding agency Colciencias, Bernal explained the project's relevance by protesting the impending disappearance of isolated indigenous populations:

> It is no mystery that the populations that inhabited our continent before Columbus's arrival have gradually become extinct, whether through acculturation and incorporation to cities and towns, or due to the high morbidity

and mortality of infectious diseases and malnutrition that followed the disruption of their habitat wrought by "white" colonos. The medical and genetic study of these populations is therefore urgent, and this becomes even clearer when we consider that indigenous settlements in other parts of the continent have been the object of these kinds of studies for over twenty years and meanwhile nothing similar has been carried out in Colombia. (Bernal 1991a: 3)

This sort of salvage genetics is heir to the anthropological anxiety of the middle of the twentieth century, which was concerned with the seemingly evident disappearance of traditional indigenous groups.[13] This anxiety was the driving force behind countless salvage ethnographies that, in the name of science and humanity, sought to create a register of those populations that allegedly were disappearing. In anthropology this discourse was deeply questioned, but in the Human Expedition it seemed to reemerge with a new face and in a different register as an argument substantiating the need for clinical and population genetics research. Just as in the 1940s anthropologists saw modernization as the culprit of the loss of traditional ways of life, for the scientists of the Human Expedition the homogenizing effects of globalization played an analogous role in killing cultural diversity:

In the first case we will reflect upon the conceptual wealth in each visited community in order to understand the importance of cultural diversity in a society that is subjected to homogenizing pressures such as those posed by the means of social communication that we call the Internet. It is possible that, in a few years, the advances of technology will place this medium in the hands of the majority of the planet's population, in a similar way to what has occurred with radio and television. When this happens, the great diversity of the Earth will be reduced to a few displays in museums that will show how in earlier times people were very different from one another and how this difference enriched humanity like colors and sounds enrich the landscape. This sad futuristic prediction fills us with the necessary enthusiasm to continue along the path that was traced by the Human Expedition, which is to collect elements that highlight the diversity that still exists today. (Gómez Gutiérrez 1998: 145)

As anthropological studies of globalization have shown (cf. Inda and Rosaldo 2002), things are much more complex than is suggested by these apocalyptic readings of the ineluctable threat of cultural homogenization. Nonetheless, these kinds of representations of imminent danger dovetailed perfectly with the arguments that justified the Human Expedition program's existence: "This is how, aided by the numerous eyes of experts of many disciplines as well as students in training, this last expedition of the twentieth century seeks to describe the characteristics that make these isolated communities attractive not only to the taxonomist or the brainy anthropology student but to any human being" (Gómez Gutiérrez 1998: 27).

In their view, Colombia's human diversity was predominantly found in those remote places where "unknown Colombians" lived. "The search for the hidden Americas," then, meant undertaking the work necessary to reveal that hidden Colombia, the Colombia of "isolated communities," of "inaccessible geographies," of "remote times." During the closing event of the GHE, held on 27 September 1993, Bernal stated, "The Human Expedition has transcended more barriers in order to put us in contact with the other Colombia, the Colombia of Colombians that we don't know, the Colombia that travels by foot, by mule, or by boat, the one that cannot know airplanes, and for whom the only experience of state presence is a teacher who shows up and doesn't stick around" (1993: 155). "The search for the hidden Americas" denotes a tone of scientific discovery of the realities that had been kept hidden and that required the mediation of the scientist's expert knowledge to surface and be recognized.

Within this framework, diversity was understood to exist predominantly in "isolated communities" that could only be accessed by crossing long distances and overcoming the most varied vicissitudes and adventures. In a place other than the lab, in the antipode of spaces and peoples transformed by civilization, is where we could find those isolated communities, which were seen as both the source (the constituent elements) and the prior historical moment of "genetic and cultural mestizaje." Gómez described this as Bernal's effort "to bring together a greater and greater number of research initiatives centered on the main premise of leaving the classroom in order to arrive at the remote places where thousands of people, who have been unable or unwilling to integrate into dominant civilization, have found refuge; and who hold the source of our genetic and cultural mestizaje" (Gómez Gutiérrez 1998: 133–134).

In these remote geographies, which signal the existence of unknown peoples, one can decipher the clues of the real country, the "deep Colombia." Although this sounds similar to Bonfil Batalla's (1996) well-known notion of a "deep Mexico," which he contrasts to an "imaginary" Mexico that is modeled on European modernity, the two concepts are not analogous. For Bonfil Batalla, the indigenous cultural grid is the source of true Mexicanness, and all Mexicans possess it even though it is generally implicit, lying inadvertently and constantly suppressed. For the Human Expedition, "deep Colombia" is not lodged within all Colombians, nor is it at the heart of a mestizo Colombian-ness. Rather, it is another, historically prior and therefore static nation, which continues to stand on the margins of mainstream Colombia, the modern Euro-Andean nation that, due to its dominance, does not need to be investigated and described. According to the academic vice president of the Universidad Javeriana, who made the following statement during the Human Expedition program's most effervescent moment, this project sought "the rediscovery of the contemporary national self": "Today, sixty researchers belonging to the most diverse set of disciplines scatter across the fragmented map of Colombia in order to sketch its genetic, social, cultural, political, and economic reality. The expedition visits the most remote places, makes a map, compares it to others, integrates it and publishes it. In this way, a new map of the true country, its living society, is being constructed" (Sanín 1992: 7–8). At the height of the program's activities, a national newspaper described the project using the language of "the other nation," the hidden Colombia. The article, via the Human Expedition's own discourse, portrayed a nation with an extraordinary ethnic wealth, which could be found in those remote rural areas that are closer to nature and have remained environmentally and morally uncontaminated. In brief, it presents us with a rearticulation of the discourses of the noble savage and pristine nature:

Colombia is not only the country of the *paisas*, the *costeños*, or the *cachacos*.[14] Nor is it the land where environmental pollution, indifference, and intolerance reign. It is also the second richest country in ethnic diversity in the world, with more than eighty ethnic groups and communities of African, Asian, and European origin. Many of them make up the other nation, the one that is used to traveling by foot, by mule, by motorboat; to laying their bare feet on the soil, feeling it, and therefore caring for it. The one that loves nature, the ocean, water. . . . In that other nation there are places that have not yet been polluted. In fact, they make the great landscape, the main reason to live amid pure air, nature, and its gifts. There are towns where envy and deceit are unknown; where people live calmly and work hard. Although they suffer, because few people outside of their communities are aware of their existence . . . Expedición Humana, their project, has been taken up by the Universidad Javeriana and its director geneticist Jaime Bernal Villegas. Its purpose is, precisely, to rediscover and raise awareness about that other Colombia. (El Tiempo, 1 February 1993)

The emphasis on isolated communities went beyond the Human Expedition's national interests and resonated with the project's international partners. The Human Expedition's joint project with the Academia Real de las Ciencias Exactas Físicas y Naturales in Madrid, for example, also "centered on the detection and study of biological markers" in "isolated populations" in Mexico, Venezuela, Chile, Argentina, and Colombia (*Boletín Informativo Expedición Humana 1992*, July 1989, 4). This joint research project continued until 1994, suggesting that the attention to isolated communities—understood as expressions of biological diversity—in defining units of analysis that explained particular national formations was not a particularity of the Human Expedition program.[15] In the end, what was at stake in the program was the strengthening of national identity through the molecular examination of the Colombian population (Gómez Gutiérrez 1998: 201). From the geneticist's perspective, this meant mapping the genetic composition of the Colombian population, even if initially this focused on "isolated communities." Gómez pointed out that "one of the paths outlined by the Human Expedition's director was to produce the genetic map of Colombia." For him, "this meant that, taking a sufficient number of representative samples from various national ethnic groups, one could hypothetically construct a global chart of the genetic content of our population" (Gómez Gutiérrez 1998: 148).

For the Human Expedition's geneticists, the recognition and valorization of the Colombian population's enormous diversity was a key way to strengthen national identity. During an interview that Nina S. de Friedemann and Diógenes Fajardo conducted with Jaime Bernal in 1993, Bernal made this quite explicit:

NSF: Is the search for diversity the main justification, the point of departure for the Human Expedition?

JB: Yes, this is how it was born. In order to observe Colombia's amazing diversity, the second richest one in the world. Here [in Colombia] the geneticist has an exceptional paradise at his disposal. This is due to its ethnic variety and the possibility of understanding it, seeing it, perceiving it from various angles that range from its visual appearance to its genetic structure. . . . In essence, we geneticists are searching for the cause of diversity. Why are we different? How does it make sense for us to be different? What are those differences and how do they impact our ways of living? (Friedemann and Fajardo 1993: 211)

The second most diverse country in the world, the geneticist's paradise due to its ethnic variety, which is expressed on the body and is visible to the naked eye or observable in the genes and therefore legible only to the expert—diversity,

Figure 2.1. The Human Expedition's logo.

in Bernal's view, is a fact that must be explained through the deployment of genetic knowledge, even if not through it alone. In his account of the Colombian nation, diversity is glossed over as "ethnic," and "indigenous communities" and "black groups" are its embodiment par excellence. Black and indigenous populations were made synonymous with isolated communities, which functioned as the program's main referents of diversity. This conflation, which was implicit in many cases, is crystal clear in a quote from Gómez: "The Human Expedition frequently travels to meet with our country's isolated populations, mainly black and Amerindian ones" (Gómez 1992: 10).

Up until the late 1980s, the notion of human diversity was also easily translated into a racialized classification of blacks, mestizos, and indigenous people.[16] Thus, for example, the Human Expedition's logo (see figure 2.1) is an allusion to this racialization of difference. We see the profiles of three men (not women, not children) next to one another in order to highlight certain somatic markers (such as hair texture, the shape of the nose, the eyes, the lips). The different figures are intended to represent, somehow "obviously," an indigenous face, a white one, and a black one. The logo's racialization is occasionally made explicit: "This project [the American Biological Bank] sets out to provide a safe storage place for the representative organic sample of the *three races* that make up this program" (Gómez 1991: 8, emphasis added).

The logo was first used in July 1989 as a header for the second issue of the program's newsletter, *Boletín Expedición Humana 1992*. Its creation was announced in the following way: "At the head of this newsletter is the expedition's new logo, which was kindly designed by Maestro Antonio Grass, who captured and represented the central idea of the project in a way that is simply unbeatable" (p. 4). The Human Expedition logo was created before the 1991 Constitution's multicultural effect was felt, and it was conserved throughout the duration of the program and in publications such as *América Negra* up until the journal's disappearance in 1998.

The role of race in the categories used by the GHE is illustrated further in a publication titled *Demographic Aspects of Indigenous, Black, and Isolated Populations Visited by the Great Human Expedition*, in which three expedicionarios make reference to the methodology that was used to collect field data. For the purposes of the GHE, a survey that operated as a genetic research protocol was designed. Mendoza, Zarante, and Valbuena state that the research instruments were the result of "meetings with various working groups of the participating research projects so that the questionnaire that was filled out during each visit could adequately respond to the needs of each group" (1997: 23). These surveys, which were accompanied by sample collection, asked for respondents' names, age, sex, and place of origin, which should be specified by "indicating the community, population, or municipality that they came from" (25). Interestingly, they also asked for the respondents' racial group: "Racial Group: In this section we identified the group to which each individual belongs, whether *indígena*, *mestizo*, *negro*, or *colono*. All individuals who had an indigenous ancestor in the first degree of consanguinity were classified as *mestizos*, and individuals who had no knowledge or documentation of indigenous relatives were classified as *colonos*" (25). What stands out here is the use of the category "race," with the inclusion of colono as a racial group, defined apparently by migrant status (settler or colonist), but in practice defined by the perceived absence of indigenous ancestry and thus implicitly white. The ambiguity of the notion of race is evident here, alongside its profound roots in notions of ancestry.

Yunis and the Regionalization of Race

As we have mentioned, in the 1990s other population genetics research projects were under way in Colombia. Most notably, Emilio Yunis Turbay, one of the founding figures of genetics in Colombia, carried out a study of more than

60,000 individuals who had paternity tests done at the ICBF (Colombian Institute of Family Welfare) between 1975 and 1992. Taking an individual's birthplace as a signal of geographic origin, this study analyzed "eight genetic systems that group 23 allelic genes, all of them blood groups distributed across the national territory according to the origin of each individual" (Yunis Turbay 2009: 94).

These studies focused on tracing the mestizaje that had taken place in Colombia, showing how the proportionality of mixture changed in each of the country's regions: "The genetic nonhomogeneity of the Colombian population is evident, as is the patchy distribution of different ethnic components, which shows beyond any doubt the existence of regions that can be differentiated by their genetic-racial composition. These regions have been differentially and racially valued" (Yunis Turbay 2009: 94). This is a view of Colombia as a "racial mosaic" and signals a clear regionalization of race. For Yunis, these studies "showed a mosaic-like distribution of [genetic] contributions, which can only be interpreted as the result of an exclusive and oppressive process of admixture (mestizaje) that was anything but open and spontaneous." And this pattern was interpreted as evidence that "mestizaje was a 'regionalization of race'" (Yunis Turbay 2009: 312).

In his book ¿Por qué somos así? (Why Are We Like This?), Yunis presents a set of maps that give an account of the "genetic structure" of the Colombian population. These maps, which are divided according to each of the five "natural regions," show the percentages of black, Caucasian, and indigenous components in each region of this mixed Colombia. Yunis argues that "the Colombian mosaic, in terms of the genetic contribution of the three ethnic groups considered—black, indigenous and Caucasian—takes on its clearest expression when we consider its contribution to the different political regions in which the country is divided" (Yunis Turbay 2009: 88–89). In a prior book titled ¡Somos así! (This Is What We Are Like!) he stresses, "Colombia is a genetic mosaic as a result of a selective process of mestizaje and regionalization of race. And for this reason in the country we have black areas, areas of mulato preference, of indigenous mestizo predominance, areas that have represented themselves as white, alleging a pretense of racial purity" (Yunis Turbay 2006: 289).

In a chapter titled "On the Origin of the Colombian Population," cowritten with his son José Yunis Londoño, Yunis makes reference to new research in population genetics. Specifically, he details studies of Y chromosome haplotypes and mitochondrial DNA, not only to reaffirm his argument that the genetic components that make up mestizo populations vary depending on geography, but also to explain how maternal and paternal ancestries differ. When referring to the paternal ancestries of "Amerindian," "Afrodescendant" (from the Chocó region), and "Caucasoid" populations of the Andean region, Yunis identifies very little Caucasoid influence for the first two. Meanwhile, he states that in the "Caucasoid population of the Andean region" there is "a predominance of European Y chromosome haplotypes, mainly of Spanish origin, and which correspond with the regions of Andalucía, Castilla, and Extremadura, while the contribution of Afrodescendant and Amerindian lineages is very low" (Yunis Turbay 2006: 271). Regarding the components of the "Amerindian population," he states that "there is a very small contribution of Caucasoid and Afrodescendant populations." And for the Afro-Colombian population of the Chocó, "the findings show that paternal lineages have been conserved as African-origin lineages, with very small contributions of Caucasoid and Amerindian populations" (271). In sum, the markers of paternal ancestry of these different "populations" differ considerably.

Based on mitochondrial DNA analysis, Yunis (2006: 288–290) identifies the frequency of Amerindian mitochondrial haplogroups in "mestizo populations" in eleven departments as well as "mitochondrial haplogroup L, which identifies African-origin mtDNA" in the same departments.[17] He concludes "that there is a predominance of the Amerindian trace in all of the regions of Colombia, ranging from 73.9 percent to 96.5 percent, with a general average of 85.5 percent, which means that the mitochondrial DNA transmitted by present-day Colombian mothers is mostly Amerindian. The contribution of other mothers is minimal" (289). Regarding the component percentages found in the African-origin mitochondrial haplogroup, Yunis remarks, "Clearly the arrival and imposition of slavery is of great significance, again, because of the selective and oppressive process of mestizaje that it established in the black population. This is the reason why the contribution of black mothers, via mitochondrial DNA, is reflected in the corresponding graphic, which shows—from the perspective of molecular transmission from mother to child—regional differences that are evident to anyone that knows the country in even the most basic way. These results must be interpreted in relation to the existence of [gold] mines, palenques, zambo populations, among others" (289).[18]

In the narratives that we have examined, it is clear that Yunis conceives of the relationship between nation and difference through the notion of race and the idea that it is regionalized. He argues that there exist "mestizo populations" that vary geographically according to the different proportions of the three ethnic components or races (Caucasian/European, black/African, indigenous/Amerindian) that historically inhabited each region. From this, he elaborates a notion of "Caucasian mestizos" or "mestizo-Caucasian population" (e.g., Yunis Turbay 2009: 130, 131). However, it is not clear whether the Amerindian and Afrodescendant categories that Yunis identifies also fit into his differentiation of mestizo populations.[19] According to Yunis, the nation's mestizo populations are spatially differentiable. Regions and departments, for example, are the embodiment of the historical processes whereby differentiable racial markers became emplaced. Moreover, the histories of each population's ancestry markers are gendered. For example, while the presence of European markers is overwhelmingly associated with fathers, and mothers' markers are mostly indigenous and African, these numbers change from place to place. Thus, the gendered composition of ancestry varies as we shift our gaze from "Colombian Amerindian populations" to "Afrodescendant populations from Chocó."[20]

Violence and Forensic Genetics

Nowadays, many of the university laboratories that do population genetics or genetic anthropology support their research with work that they carry out in medical and, now most prominently, in forensic disciplines. This shift toward forensic genetics in Colombia is the result of the escalation of the armed conflict at the end of the twentieth century. Although in the last decade human genetics research in Colombia has changed substantially, in both its orientation and technology,[21] the regionalized grammars of difference that Yunis (2006) depicts, and some of the ideas about isolated communities born in the GHE, continue to shape its horizons.

At the dawn of the twenty-first century, initiatives such as Plan Colombia ignited a renewed interest in forensic science, resulting in the arrival of new standards and technologies of genetic research in order to identify victims and perpetrators of violent acts.[22] In this scenario, individualized forensic identification has been a driving motivation of the search for molecular differences among Colombians.[23]

The famous CODIS (Combined DNA Index System) used by the FBI in the United States, which integrates DNA technologies such as microsatellite repetitions to identify bodies, was one of the first scientific contributions of Plan Colombia. Within this forensic field, Manuel Paredes, a student of Yunis, was the first geneticist to create a forensic lab for criminology purposes inside the Instituto de Medicina Legal (Legal Medicine Institute) in 1993. Also, expedicionarios such as Bernal and Zarante helped create a forensic genetics lab within the national police and trained the first generation of criminal investigators.

Emilio Yunis continued his work on forensic genetics at the ICBF until the beginning of the twenty-first century and then continued doing forensic genetics work for his own private laboratory. As a consequence of the volume of genetic tests needed for both civil and criminal cases, and the public importance of such work, forensic genetics has become a new site for competition and dispute among genetic experts (see Schwartz-Marín et al., 2013). However, without a doubt the work and standards produced by the genetics laboratory at the Colombian Institute of Legal Medicine have had the upper hand, becoming the referent against which all other identification processes in the country are measured.

In Colombia, the CODIS has been tailored to identify human remains, suspects, and victims according to their geographic origins in four populations: two populations of African descendants from the Pacific and Caribbean coasts, mestizos from the Andean region, including the Amazonian and Orinoquian regions, and a southwest Andean region with an "important Amerindian component" (Paredes et al. 2003: 68). Although the DNA markers used by the CODIS database are the same regardless of country or forensic system, the boundaries of what legitimately constitutes a population differ from place to place. What is striking for Colombia is the production of genetic difference by forensic disciplines in a way that emphasizes the notion of a country of racialized regions, which are apparently clearly distinguishable in terms of allelic frequencies.

The dominant investigations in population genetics, many of them in the last ten years done from a forensic perspective, deepen the national common sense of a racialized and regionalized Colombia.[24] Moreover, these imaginaries have become naturalized and standardized as a tool to order the whole range of forensic genetic inquiries in the country. In practice this means that every time the forensic system receives a new sample for analysis, it is classified according to the four racialized populations described in Paredes et al. (2003). Thus the Colombian forensic system can be seen as a machinery in which the imaginaries of difference that have been historically intertwined with genetic science become reproduced and reinforced with unprecedented scale and public importance.

Conclusion

While the Human Expedition looked at isolated communities in order to draft the genetic map of the Colombian population, the working group led by Yunis was carrying out the "first great study of mestizaje in Colombia" (Yunis Turbay

2009: 94), emphasizing the varying percentages of black, Caucasian, and indigenous components according to geographic divisions (regions), political administrative units (departments), and other spatialized historical processes (e.g., colonization). In both cases, they were investigating molecular diversity, a particular kind of diversity that cannot be perceived by the naked eye and therefore requires the competence of genetic and genomic knowledge and technology. For them, isolated communities and the spatialized variations of mestizo populations were the privileged sites of analysis to understand molecular diversity and the building blocks with which the Colombian nation was imagined.

Despite their differences, in both cases national molecular diversity was racialized and spatialized. On the one hand, in the work of Yunis and his collaborators the use of the term "race" and its regionalization are explicit. His work, first with blood markers and later with ancestry markers (haplogroups) in mitochondrial and Y chromosome DNA, operates with a racialized logic that is transparent in the very terms that are used to describe difference: Caucasian/European, black/African, indigenous/Amerindian. On the other hand, in the Human Expedition program, although there were occasional explicit references to race, some of the project's participants now publicly reject the concept (Gómez Gutiérrez, Briceño Balcázar, and Bernal Villegas, 2011).[25] Nonetheless, as we have shown, there are clear instances of the open mention of three races and implicit but obvious references to race, as in the program's logo. Given the Human Expedition's particular history and affiliations (to Nina S. de Friedemann, to name one), in its publications and internal documents we can see the burgeoning discourse of multiculturalism being used much more often, replacing the term "race" with more politically correct references to "culture" and "ethnicity." We are not arguing that the Human Expedition or Yunis engaged in genetic reductionism, a practice that they have explicitly and repeatedly opposed. Nor do we attribute to them the kind of racial thinking that was characteristic of the early twentieth century, which established direct correlations between a population's biological characteristics and its specific behaviors and intellectual or moral capacities. However, if we search for processes of racialization that do not require the explicit iteration of the term "race," we can see that the concept of race appears every time that discrete categories of difference are mobilized and a distinction is made between indígenas, negros, and mestizos. In this sense we can state that there is a clear racialized articulation of difference, independently of whether the word "race" is mentioned.

The spatialization of difference is also a central trope in the genetic imagination of the Colombian nation. In the Human Expedition program, the very idea of an expedition—that is, the pressing need to travel to remote places that remain hidden and must urgently be discovered by the gaze of the scientist—evidences a geographic imaginary of difference. Likewise, for Yunis and his collaborators, mestizo populations are not all the same, given that they have been historically marked by a fragmented geography of natural regions that has resulted in a clear spatialization of racialized difference.

Notes

1 The historical narrative of this chapter was put together through the accounts of Colombian geneticists in Medellín and Bogotá in the following universities: Universidad de los Andes, Universidad Javeriana, Universidad Nacional, and Universidad de Antioquia. In addition, we incorporated the accounts of geneticists who work in state institutions such as the Instituto de Medicina Legal, Instituto Colombiano de Bienestar Familiar, and the National Police.

2 The work of Dr. Hugo Hoenigsberg (from Uniandes) and Dr. Margarita Zuleta (from Universidad de Antioquia) actually predates Emilio Yunis's pioneering studies of clinical and population genetics in Colombia. However, neither of them initially did human genetics, instead focusing on fruit flies. Margarita Zuleta had studied population genetics with Hermann Muller (Nobel Prize, 1946), and, according to Dr. Gabriel Bedoya, she could be considered one of the first population geneticists in Colombia.

3 The lab's web page used this wording in 2012 (accessed 28 October), but it has since been changed to refer to "different Colombian populations" (Laboratorio de Genética Humana 2013).

4 The Human Expedition program, however, did not establish direct ties with either the Human Genome Project or the Human Genome Diversity Project. In an interview, Jaime Bernal emphatically stated, "There is absolutely no relation with the HGP or the HGDP . . . not only do I have nothing to do with them, but I have consciously avoided having anything to do with them; before the debate began, at a geneticists' meeting in Rio de Janeiro a few years ago, they were about to name a commission of Latin American geneticists to be on the board of the Human Genome Diversity Project and I left the conference because I knew that if I stayed I would be named to be on that commission and I was not interested in participating in that" (interview cited by Ramos 2004: 15).

5 A publication from 1994 lists the "national and foreign institutions" with which the institute collaborated (Bernal and Tamayo 1994: 39, 41). The foreign ones include institutions in the United States, France, Italy, Portugal, Spain, Scotland, and England. One of the researchers of the Human Expedition program, Genoveva Keyeux, was a member of the UN's bioethics committee.

6 In fact, one of the volumes of this collection is titled *Variación biológica y cultural en*

Colombia (Biological and Cultural Variation in Colombia) and presents the most direct results of the Human Expedition program in twenty-nine chapters that cover a very broad range of themes, some of which had already been published in other places such as the journal *América Negra*, which was the expedition's official publication (Ordóñez Vásquez 2000). The Colombian Institute of Anthropology and History also participated in the collection's design. This institute is the government entity charged with carrying out research projects related to indigenous groups and administering the nation's "archaeological patrimony."

7   This same wording can be found in several texts: see for example Bernal and Tamayo (1994: 31).

8   This is reminiscent of the notion of marginality, which is culturally and politically constructed in opposition to ideas of an urban, mainstream self. The severance from the mainstream, Tsing points out, is not a flat isolation, but rather a heterogeneous process that is "the source of both constraint and creativity" (1993: 18).

9   The university still keeps a copy of a document by Ignacio Zarante titled "Personal Equipment to Travel on a Great Expedition" (Instituto de Genética Humana, Universidad Javeriana, http://www.javeriana.edu.co/Humana/equipo.html). The document reveals something about how these trips were imagined and undertaken.

10   For a thorough description and analysis of this debate, see Ramos (2004), Uribe (2010), and Barragán (2011).

11   The sudden death of Nina S. de Friedemann, in October 1998, interrupted the publication of *América Negra*.

12   This is the term that was used to refer to these populations up until the 1990s. After this, they were increasingly referred to as Afro-Colombians.

13   Of course, salvage anthropological efforts have a long history that precedes anthropology as a discipline and is heir to the simultaneous destruction and fascination of the Other that was wrought by colonial encounters. Although the term was coined in the 1960s as a critique of colonial practices within the discipline, the logic of salvage anthropology has continued to motivate the collection of cultural (and in this case biological) elements that are perceived to be threatened by the advance of Western civilization. As critical scholars note, its logic continues to reverberate both across time and space (cf. Stephens 1995).

14   These terms are used to refer to people from different regions. Put simply, *paisas* are from the department of Antioquia and its surrounding areas; *costeños* are from the Caribbean coast; and *cachacos* are from Bogotá.

15   In the April 1990 issue of the Human Expedition's newsletter, *Boletín Expedición Humana 1992*, there is an article titled "Génesis biológica de las nacionalidades hispanoamericanas" (Biological Genesis of Hispano-American Nationalities), which states that the joint research project's goal "is a collaborative effort among several Latin American centers," the goal of which is "to give a general idea of the genetic characteristics of our people" (p. 8). The journal *América Negra* (June 1992) published a list of research projects affiliated with the Human Expedition, which

included a project called Biological Genesis of Hispano-American Nationalities, overseen by Jaime Bernal and the Real Academia de Ciencias Físicas y Naturales de España.

16 This classification is racialized even if the term "race" is not explicitly used and despite the insistence that biological race does not exist. It is racialized because it uses historically racialized notions such as *negro*, *indígena*, and *blanco* (or its substitutes such as Caucasian, African, European, and Amerindian) as referents to think about the country's cultural and biological difference.

17 In Colombia, departments are the largest political-administrative units. Chocó is not included among these departments, but rather Meta, Cundinamarca, Boyacá, Nariño, Santander, Norte de Santander, Tolima, Valle del Cauca, Córdoba, Sucre, Atlántico, and Antioquia.

18 Yunis is implicitly referring to the historical-geographic patterns of slave labor in Colombia, which centered on gold mining in the Pacific River basin. He is also referring to slave runaway communities or *palenques* and processes of indigenous-black miscegenation in this region that produced "*zambo* populations."

19 From the maps Yunis presents, it is clear that the population of the Pacific region, including the department of Chocó, has indigenous and Caucasian genetic components, much like mestizo populations in general (Yunis Turbay 2009: 360–366).

20 This view echoes two notions that have long histories in Colombia. The first is that Colombia is a nation of regions. The second is that one of the most salient differences between these regions is their racial composition. Together, these ideas amount to a persistent regionalization of race and a racialization of (naturalized) regions, which are heir to environmental determinism, biological racism, and national exceptionalism and have been recurrent, if dynamic, components of the imaginaries of Colombian uniqueness.

21 For a discussion of this period of genetic research in Colombia see chapter 5 in this volume and Olarte Sierra and Díaz del Castillo (2013).

22 Plan Colombia is a bilateral agreement between the U.S. and Colombian governments, which began in 2000 and still exists today. Plan Colombia provides U.S. funding to support the military in the Colombian state's twin internal wars: "the war on drugs" and "the war on terrorism."

23 Although paternity tests were being carried out in the 1970s, it was not until the 1990s that the Colombian state's concern with the escalating war fostered the development of new forensic genetic technologies. Initially, following the passage of Law 75 in 1968, blood groups were employed in paternity cases; in the 1980s new technologies such as HLA began to be used.

24 A large number of these research projects are master's theses from biology or genetics programs at the Universidad Nacional, Universidad Javeriana, Universidad de Antioquia, Universidad del Valle, and Universidad de los Andes. Some of them (e.g., Díaz 2010; Terreros 2010) have been financed by the Instituto Nacional de Medicina Legal y Ciencias Forenses.

25 For example, during the presentation of preliminary findings of the research project on which this book is based, held at the Universidad Javeriana on 9 December 2010, Alberto Gómez was emphatic in questioning the relevance of the term
"race," thereby distancing genetic research from any form of racialist reductionism. See also Gómez Gutiérrez (2011).

chapter 3

Negotiating the Mexican Mestizo

On the Possibility of a National Genomics

Carlos López Beltrán, Vivette García Deister, and Mariana Rios Sandoval

The aim of this chapter is to provide a historical and analytical framework to situate and better understand the episode of the Mexican Genome Diversity Project (commonly described as the Mexican Mestizo Genome Project) developed by Mexico's National Institute of Genomic Medicine (INMEGEN), and peculiarly framed in a racialized matrix. We explore the nationalist rhetoric in which it was clothed, and the scientific and political reasons for its inadequacies and shortcomings. One central issue we address is INMEGEN's focus on the figure of the Mexican mestizo as an icon for the national population it was investigating.

Since the nineteenth century, the mestizo in Mexico has been a powerful, complex, deeply rooted, ideological and identitary construct at the center of a successful nation-building project. It has provided an egalitarian reference for el mexicano with which to conjure racial and ethnic heterogeneity, and disruptive economic and political difference. As with many aspects of Mexican modernity, there is an increasing perception that the mestizo as an axis of Mexican identity is in crisis. After many recent demographic and political transformations, it is no longer efficiently gluing together national identities, and the negative aspects of its intended racial homogenization are becoming more relevant, according to a number of authors (Aguilar Rivera 2001; Lomnitz 2010c; Tenorio Trillo 2006; Viqueira 2010). In what follows we sketch a political and ideological story of the mestizo, and then concentrate on the way it has been historically configured within the bounds of anthropological and biomedical pursuits.

We are interested in the ways in which national and international scientific communities have staked their claims on the figure of the Mexican mestizo, each with its peculiar bias. We begin with a brief historical recapitulation of the trajectory of the mestizo's rise to iconic status as a locus of national identity and then turn to a special moment in this history, when the mestizo was transformed into a scientific object within anthropology and biomedicine. The succession of biotechnologies and racialized markers from serology, immunology, genetics, and more recently genomics have consistently produced differentiated descriptions of the Mexican *indio* and the Mexican mestizo. Our aim is to clarify how and why genetics and genomics have of late revisited the mestizo, recasting the figure in the most up-to-date scientific molds by allotting to it particular percentages of Amerindian, European, and African genomic ancestry. Through the lens of its latest iteration as a unit of genomic analysis, we aim to understand the continuities and ruptures in relation to analogous inquiries in the past, especially in the second half of the twentieth century, and reveal some of the workings of the interface between science and ideological frames.

The Mestizo as an Ideological National Icon

A detailed history of *mestizaje* and of the mestizo in Mexico has yet to be written. We have a fragmentary account of this story (Aguirre Beltrán 1946; Alberro 2006; Basave Benítez 1992), but we lack the larger narrative. Many historians have sketched a basic outline, consisting of a series of milestones and historic nodes that would need to be strung together into a complete interpretation. Among these formative elements, the following have set the parameters of a scholarly historical paradigm: the unequal sexual reproductive encounters during the European invasion (Salas 1960); the enslavement of African peoples and their emplacement in New Spain (Aguirre Beltrán 1946); the development of a colonial *sociedad de castas* and its progressive disruption during the Bourbon period of the eighteenth century (Chong 2008; Katzew 2005); and the long period of different evaluations of (and solutions for) mestizaje, after Independence (Teresa de Mier 1987 [1811]; Tenorio Trillo 2006).

Such a story cannot assume invariant relationships between race, mestizaje, and social mobility in different historical periods. But the components of the so-called pigmentocracy that were described by Humboldt in 1822 were reconfigured to generate Eurocentric or Indocentric versions of *mestizofilia*, a legitimating ideology forged in nineteenth-century civil and external wars (Basave Benítez 1992; Izquierdo and Sánchez Díaz de Rivera 2011; Sánchez-Guillermo 2007).

The Viceregal or colonial period (sixteenth–eighteenth centuries) saw the emergence of a structuring principle that organized the enormous variety of the *castas*, or racial admixtures, and imposed itself upon the social and political scene. There were basic first-order castas (mestizo, mulatto, and *lobo* —the term ascribed to a person of African and indigenous parentage) and also second-order sliding and unstable castas, labeled by scores of derogative names. Within the order as a whole, a normative whitening trajectory operated, positing the Spanish physical type as the ideal. However, other irregular and seemingly random trajectories also emerged during this period, finding their exemplifications in the tradition of casta paintings. The nonwhitened, unruly casta bodies are perceived in these paintings as Indianized or Africanized and they are labeled with a rich vernacular of degrading names, such as *coyote*, *zambaigo*, or *chino* (López Beltrán 2007, 2008). The racialized categories of mestizaje in Mexico began to stabilize toward the end of the Viceregal period within the following main categories: Spaniards—or, for those born in America, *criollos*—mestizos, indios, and castas, or all the other unclassifiable appearances or types (Aguirre Beltrán 1946).

The post-Independence epoch, in spite of the abolition of the casta system and slavery, did not witness the destruction of the ideological power of the European racial civilizing matrix, but rather its redeployment, as one of the major divides in nineteenth-century Mexican culture opened up between those who wanted to preserve the white European body as the regulatory ideal for migration and demographic policies and those who believed it was wiser to move this ideal toward a body and complexion that looked like the Mexican majority. In other words, the criollo type was positioned against the mythic original mestizo, offspring of the indio and the Spaniard (Falcón 1996).

Almost everyone in Mexican policymaking circles in the nineteenth century agreed that racial homogenization was essential to the survival of Mexico. But the ideal racial type of the model Mexican citizen in the new nation, and the policy formula that would concoct it, were a source of deep dissent. A criollo majority program was radically different in orientation from the policy to increase mestizaje, especially with regard to the originary Indian groups. One strategy was to isolate the indios and leave them to a Darwinian process of extinction in the struggle for survival; meanwhile, a policy of selective European migration would improve and whiten the existing mestizo base.

Taking up Basave's well-known notion, we can call this strategy white mestizofilia. Another strategy, which we could call swarthy (*morena*) mestizofilia, took a more tolerant view of the indigenous groups and aimed at incorporating them into the main mestizo body in all parts of the country, by means of racial fusion and acculturation. Liberal factions associated themselves with this second strategy, and it was probably the only realistic one, due to the fact that European mass immigration, such as that which occurred in the United States and Argentina, never materialized in Mexico (Lomnitz 2010b; Saade Granados 2009).

By default, the ideal Mexican body came to be identified with the swarthy (moreno) mestizo as a result of the fact that the anti-Spanish *liberales* (the liberal party) won their political struggle against the pro-Spanish *conservadores* (the conservative party) and used the idealized image of the swarthy mestizo (in distinction to the paler, Europeanized whitened mestizo) to generate a national ideology organized around mestizofilia, making the darker mestizo the emblem of racial, cultural, and political unity (Basave Benítez 1992; Stern 2000). The putatively all-inclusive category of mestizo was enthusiastically adopted by the ideologues of the Mexican Revolution.

Still, there were drawbacks to the mestizo as a nationalist identity figure. For one thing, it allowed no legitimate space in Mexico for foreigners and individuals of different ethnic backgrounds, such as Jews and Asians (Lomnitz 2010a, 2010b). It tolerated "otherness" only at the extremes of the population mix: the white Europeans and the Mexican indios (Lara 1990; Renique 2003). However, while white Europeans had the political and economic capital to protect themselves, the planned and "inevitable" eradication of indios through acculturation and assimilation into the mestizo showed the dark side of the national project of mestizofilia. In fact, postrevolutionary *indigenista* policy made the mestizo/indio binary an arena in which the indios could be converted into mestizos (Navarrete 2004). Crucial to this effort were education and sanitation programs designed to play a key role in the postrevolutionary vision of the modern Mexican society and state (Agostoni 2011; Aréchiga Córdoba 2009).

The synergy between the swarthy mestizo icon and the nationalist ideology of the first part of the twentieth century was evident in the postrevolutionary imaginary, which officially represented the national population around an indio/mestizo axis. Scientists, especially anthropologists and physicians, reinforced this dual taxonomy and the polarities it engendered in their studies of Mexico's human geography. Under leftist president Lázaro Cárdenas (1934–1940), the indigenista project was launched, with the goal of resolving once and for all the age-old *problema indio*. The project was led by eminent anthropologists who refurbished and reinforced the indio and mestizo categories in a narrower and seemingly more precise manner (Aguirre Beltrán and Pozas Arciniega 1981; Bartra 2005; Bonfil Batalla 2004; Saade Granados 2009; Villoro 1950). To this end, linguistic (Spanish versus native) and cultural criteria were privileged. In the 1960s, a young generation of Marxist-influenced Mexicans turned against this paradigm, claiming that the paternalistic and nationalistic construal of the notions of indio and mestizo were a façade, under which capitalist economic relations were penetrating the whole of the Mexican social sphere (Medina and Mora 1983; Vergara Silva 2013; Warman 1970).

The Marxist moment passed, but there has been no letup in the onslaught on the historical and contemporary role of the Mexican mestizo as an identitary icon, led by a new generation of historians, anthropologists, and cultural critics, who claim that the mestizo has outlived its ideological usefulness as a social homogenizing and stabilizing tool and who reveal the hidden racism, exclusionary policies, and undemocratic asymmetries of power that the image condones (Aguilar Rivera 2001; Gall 2007; Izquierdo and Díaz de Rivera 2011; Lomnitz 2010b; Tenorio Trillo 2006; Viqueira 2010). Against this, a number of cultural studies scholars defend multiculturalism as a more realistic and emancipatory vision of Mexican history and contemporary society (Navarrete 2004; Oehmichen 2003).

The Mestizo Is Taken Up by Science

The history of race science in the twentieth century can be understood as the history of the challenge of applying archaic concepts—namely, purity—to human societies. Every putatively objective and scientific criterion or marker that was used to pick apart the physical and hereditary race differences, be it skin coloration, skull measurements, blood types, blood proteins, gene variants, or IQ, immediately raised the question of racial admixtures (Gannett 2004; Marks 1995; Sans 2000). As race science first appeared at the beginning of the twentieth century in Europe and the United States, Mexican scientists developed the research process of determining the degree of heterogeneity of their national population in racial terms. The political points of reference, however, were distinct, given the state's defense of the mestizo image. Mexican scientists, then, started from the assumption that there must be a basic underlying mestizo uniformity due to large-scale historical miscegenation over the past five hundred years.

Not surprisingly, between 1900 and 1950, the study of contemporary indigenous groups shaped the style and content of anthropology in Mexico (Rutsch 2007). Urban and campesino (peasant) mestizo groups, on the other hand, were seldom objects of such study. Only after the late 1940s did the mestizo come into its own as a subject of academic philosophical and psychological reflection and research, in the wake of the wave of Continental theory that washed through academia. Two generations of sophisticated Mexican authors, from José Vasconcelos to Octavio Paz (including the exceptional contribution of the disciples of José Gaos and Samuel Ramos), created a genre of theoretical analysis around the physical, psychological, and ideological makeup of el mexicano. He was typically mestizo, dark skinned, witty, slothful, passionate, unstable, creative, and gloomily self-conscious. The bodily and spiritual mixture of Spanish and Indian blood produced a paradoxical set of features. Octavio Paz's *El laberinto de la soledad* can be said to be the literary distillation of this whole genre.[1]

At the same time, the turn affected physical anthropologists and physicians. From the indio, they moved to the mestizo body, seeking to frame in physical and biological terms the fruit of generations of racial admixture. The theoretical base of these scientists was not Continental theory, but instead the advances in genetic and evolutionary biology stretching from the 1930s to the 1950s, which produced a model of DNA, molecular techniques from blood serology, an understanding of such things as the histocompatibility complex (HLA) and protein biochemistry (e.g., hemoglobin), coupled with neo-Darwinian population genetics. This did not lead to the abandonment of racial categories. Instead, molecular techniques promised to provide racial markers with a scientific basis and unveil the hidden admixed biological makeup of the Mexican mestizo population.

In Mexico, the first research work around the genetic variants of different subpopulations (mestizo and Amerindian) was carried out by means of the techniques developed in the North: the use of markers and molecular reactions, and the application of serological, immunological, and electrophoretic techniques (Suárez and Barahona 2011). At the same time, the anthropological ethos was changing. The UNESCO statements on the erroneous use of human races as scientific categories are a marker of a massive intellectual shift. Race science lost its reputation as a science, and garnered one as a disguised form of racism. Funding was cut from race science research, and the scientific observation of human populations and their diversity was reformulated to reflect the rejection of the superior race ideology, or derogatory visions of mestizaje (Maio 2001; Montagu 1942).

The technical ability to mark molecular variants and polymorphisms for blood groups and hemoglobins allowed for research on fine-grained biological variation among human populations (Chadarevian 2002; Mazumdar 1995). The dynamic hypothesis of population genetics allowed researchers to test for answers to questions about the molecular history of human groups, using techniques such as chromatography and gel electrophoresis to differentiate between normal and abnormal hemoglobins. This became part of the arsenal of human population biology and medicine. It clashed with the research style of Mexican anthropology, with its resistance to any anthropology that reduced the autonomy of culture to human biological models. Only a few Mexican anthropologists wholeheartedly turned to the molecular style, and the gap was filled instead by physicians. Human geneticists, most prominently Rubén Lisker and León de Garay, began to investigate molecularly unexplored mestizo populations, looking for their mestizaje markers in terms of frequencies in blood types, molecular variants (hemoglobins), and the enzyme G6PD (glucose-6-phosphate dehydrogenate of erythrocytes) (Barahona 2009). To this list were added the immune identity antigens, which opened a new molecular window to mestizo diversity (Arellano et al. 1984; Gorodezky, Terán, and Escobar Gutiérrez 1979).

The pioneering works of León de Garay and Rubén Lisker in medical genetics (Barahona 2009; Suárez and Barahona 2011) and those of Rocío Vargas in molecular anthropology borrowed the ready-made mestizo-indio population dichotomy to frame their research questions and answers. Paradigmatically, these studies used molecular markers to quantify the European, Amerindian, and African components of the admixed groups in every region and to single out peculiar local mutations with possible pathological and ethnic importance. The phrase "Mexican mestizo population" very soon became a common designator that captured the basic idea that the Mexican population could be seen in terms of two regions of a circle. The center consisted of admixed mestizos (ideologically identified with the Mexican), farther from the center would be indigenous people who were admixed to a significantly lesser degree. Along the radius from the center to the circumference, people could be graded according to their admixture. From a dynamic point of view, the area of admixture was spreading to incorporate the whole. The scientific image confirmed popular common sense: when all the postrevolutionary institutions (the school, the hospital, the political party) were run by and for mestizos, when any random Mexican walking down a city street was almost certainly a mestizo, one would expect scientific scrutiny to confirm the presence of mestizo characteristics in the majority of the population.

In a short time the formulaic descriptor "Mexican mestizo" reproduced itself across the spectrum of those research groups concerned with the Mexican population as a whole. It became the standard category for grouping all the samples taken mainly from patients treated in hospitals and institutions of the national health system (Cerda-Flores et al. 2002; Lisker et al. 1986). Figure 3.1 shows the growth of scientific output genetically gauging the Mexican mestizo population. Most of these studies operated with low budgets, often subordinate to other medical or anthropological projects, and produced small samples. From the 1960s onward, Mexican medical geneticists and physical anthropologists steadily accumulated molecular data of mestizaje that

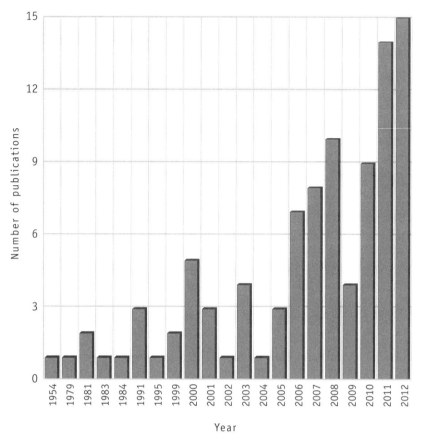

Figure 3.1. Growth of publications featuring the term "Mexican mestizo population" in title or abstract. Source: PubMed (search 13 September 2012).

converged around a set of figures giving standard percentages of European, Amerindian, and African ancestries in different regions. A number of articles appeared in specialized publications quantifying the percentage allocated to different ancestral components according to a trihybrid model of Mexican mestizaje (European, Indian, and, to a lesser extent, African). The advent of genomic sequencing and the powerful bioinformatics that came with it opened a window of opportunity to finally assess the realities of genetic mestizaje within the Mexican boundaries. As we discuss below, this task was, in the eyes of Gerardo Jiménez Sánchez, first director general of INMEGEN, both politically and technically valid.

The Genetic Study of Admixture (or "We're All Mestizos, but Some Are More Mestizo Than Others")

The project of arriving at a precise representation of the proportional genetic contribution of two or more ancestral groups (assumed to be genetically different) to a recently admixed population by means of testing samples and using standard statistical inferential procedures was taken up early in the twentieth century, when it became theoretically and experimentally possible. This avenue of research was of special interest for scientists dealing with the genetics of Latin American (typically admixed) populations. The basic idea at first was to use allele frequency data in an admixed population putatively derived from two parental populations, using a simple dilution model (Bernstein 1932; Ottensooser 1944).

The question of which were the admixed groups and which the parental ones was, of course, independently answered through reference to historical or traditional narratives outside the research frame. This let researchers assume relevant ancestral populations, as defined by historical criteria. The fact that synchronic real ancestral populations, for use as contrast groups, were nonexistent was solved by taking as a proxy some group or groups thought to descend directly from the putative ancestral one. A contemporary sample of this group would stand in as a proxy for the frequencies of genetic markers to be found in the ancestral group.

Many analytic methods were devised and used to this end. According to Salzano and Bortolini (2002), they all depended on two basic assumptions: (1) there is no factual error in the choice of parental groups or in their gene frequencies, and (2) we can discount changes in allele frequencies (e.g., genetic drift) occurring independently of assumed gene flow as statistically trivial. But one further assumption is at work in the application of this research paradigm: that the ancestor-offspring link between the putative parental groups and the putative admixed one given by historical accounts is empirically valid, which is not easy to certify within the specific research frame.

During the twentieth century, admixture analysis methods were technically improved several times, as was the actual technical access to real genetic variability. Until quite recently these inquiries depended on using only a few allelic markers, which amplified the margin of error of any conclusion. For the Mexican mestizo population, different research protocols and methods gave varying percentages of European and Amerindian contributions to the mestizo mix, both within and between the geographical regions considered (Lisker 1981; Salzano and Bortolini 2002).

With the exponential rise in computer power, genotype sequencing procedures have explosively increased the number of alleles that can be simultaneously assessed. In turn, admixture models have increased their precision, given the assumptions we outlined above. Admixture analysis features have been built into powerful bioinformatic clustering programs like STRUCTURE (Patterson, Price, and Reich 2006). But even as precision and range of the analysis have become better, the data sets still derive from the same old empirical assumptions that authorized the validity of previous admixture inferences. The basic assumptions regarding the nature of the Mexican mestizo population remain what they were in the early days of molecular anthropology and biomedicine, despite significant technological developments in both fields.

A basic background mestizo homogeneity is assumed to cover most of the national population. On top of that, the percentages of each racial parental component are seen as regionally modulated. Taken together, these premises allow for the regional heterogeneity of the mix to be contrasted with an underlying mestizo similarity. In this way, local narratives of differential admixture in each region feed on and reinforce the national narrative of Mexican racially admixed homogeneity, and vice versa. Thus, the dual view of the Mexican mestizo (as nationally homogeneous and regionally heterogeneous) seems to be the expectation (Lisker 1981).

Conceiving a National Genomics

On 19 July 2004 the federal government of Mexico announced the founding of INMEGEN, thus decreeing the official beginning of the country's national medical genomics. This announcement was the fruit of a long, concerted effort led by a small group of influential physician-politicians who came together in 1999 and resolved that Mexican biomedicine should take advantage of the recent flurry of developments on the genomic research scene. This group's consensus was that genomic research was a big science project, which should be conducted through a new research venue. From the beginning, INMEGEN's budget allocated substantial amounts to a very active campaign to promote the public image of the new institute and that of its first director, the young, well-groomed, and entrepreneurial pediatrician-geneticist Gerardo Jiménez Sánchez. This campaign launched with frequent media spots, interviews, and stories highlighting INMEGEN's plan for developing a so-called Mapa del Genoma de los Mexicanos, its flagship project. Even before INMEGEN was fully established, promoters of the institution campaigned to seal a connection between its activities and the Human Genome Project in the minds of its Mexican audience. Later, the publicity program turned to Mexican participation in the International Haplotype Map Project. INMEGEN's public image management strategy prioritized synthesizing a noticeable national agenda with a prestigious presence in international science, which was attention catching and created a rationale for INMEGEN's federal financial support.

By the time INMEGEN was created, nobody could yet claim to offer a definitive measure of the genetic components of the Mexican mestizo. Jiménez Sánchez keenly perceived this gap and, perhaps unfairly, helped launch the project to map the Mexican genome as if it were a pioneer probe into uncharted territory. Such a gesture was seen as arrogant by members of the scientific elite (especially the older generation of medical geneticists), but it was consistent with INMEGEN's program to position itself as a cutting-edge research center, independent of the North American biomedical establishment, and monopolizing genomic research among the mestizo populations south of the United States.[2] During Jiménez Sánchez's administration of INMEGEN, the institute also strove to claim as its own a population universe of study objects (and future genomic patients[3]) within preexisting political and territorial boundaries.

The flagship project of INMEGEN under Jiménez Sánchez's directorship was conducted under two assumptions regarding the characteristics of Mexican populations: (1) that they are admixed, genetically heterogeneous, and stratified, which accounts for the genetic singularity of the Mexican population; and (2) that differences in disease rates between Mexicans and other populations are due in no small part to differences in the frequencies of disease-promoting genetic variants present among Mexican mestizos. The project also relied on the hypothesis that haplotype-disease associations for common diseases would be turned up by research (Hirschhorn and Daly 2005; Jiménez-Sánchez, Childs, and Valle 2001). Because of their historically recent admixture, mestizo populations pose challenges to this type of study as potential associations could be due to ancestry (that is, population structure or stratification) and not to a disease-associated haplotype. Because genetic admixture and the resulting population stratification are known to increase the error rate in association studies, INMEGEN researchers initially perceived admixture mapping as a valuable tool for optimizing design and analysis of future genetic association studies in the Mexican population (Silva-Zolezzi et al. 2009).

First called the Genomic Structure and Haplotype Map of Mexican Population (see Schwartz-Marín and Silva-Zolezzi 2010) and later renamed the Mexican Genome Diversity Project (hence the acronym we use: MGDP), the project, in its first phase, aimed at generating the first genome-wide assessment of genetic diversity within a Latin American mestizo population. While publicizing their project as the first genotyping effort of a Latin American admixed population to make the strongest national and international media impact, the Mexican project actually started off behind similar biomedical efforts being developed simultaneously by groups in the United States (e.g., González Burchard et al. 2005; Wang et al. 2008). Even during the planning and lobbying period preceding INMEGEN, there was a latent worry among its promoters that Mexico was losing the race to sort out Mexican mestizo and indigenous samples from Mexican sources to foreign investigations; worse, there was the fear that these results might eventually lead to a loss of independence and to genomic subordination (see Schwartz-Marín 2011). As in the case of France, chronicled by Paul Rabinow (1999) in his book French DNA, a nationalist stance was critical to the early ethos of INMEGEN. The prevailing opinion was that a national institute of genomic medicine in Mexico must be empowered by a generous injection of resources in order to establish itself as an essential intermediary for geneticists from inside and outside Mexico, especially in everything related to the Mexican mestizo and indigenous populations. There was a strong view that the institute should be awarded control over any human biological sampling activity within the Mexican territory related to human genomics projects, both at the national and the international levels. In addition to the much-heralded aim of "genomic sovereignty" (a term deployed by INMEGEN's lobbyists, although it was never clearly defined), Jiménez Sánchez sounded loud warnings against "safari research" (that is, the illegal expropriation of national resources). Genomic sovereignty went from a lobbyist's bullet point to a legal concept when INMEGEN's project acquired feasibility and a position at the table of international genomics research. In a way, the term helped to dissipate some of the tensions and organize cooperation among actors with dissimilar interests, an example of what Schwartz-Marín (2011) calls "consensus without cooperation." Any collection of human biological samples within national territory carried out without authorization of the state, and their exportation to foreign laboratories for processing, was labeled by law as a potential threat to Mexico's sovereignty. Perhaps due to the publicity surrounding genomic enterprises in other countries, which have aimed at producing "national genetic databases" (see Pálsson 2007 for the case of Iceland), access to a mestizo population sample was defined as a national public good or patrimony, over which INMEGEN should be given privileged and legitimate rights.

A nationalist mestizo rhetoric, twinned with the legal and technical reinforcement of genomic sovereignty, wrapped INMEGEN's research enterprise in a patriotic cloud. But it did not quickly translate into results. Organizing actual research lagged behind political and rhetorical action; while claiming a proprietary stake in the Mexican genome, characterizing that genome turned out to be much harder than expected. Thus, as INMEGEN's scientists were advancing toward their mission, the real success of the institute could be measured in the constant presence of Jiménez Sánchez in public spaces, newspaper interviews, talk shows, celebrity profiles, and radio and TV spots. It took INMEGEN a further two years after the end of the first phase, which had focused on the so-called country-wide sampling *jornadas* or work days (2005–2007), to produce a proper publication in an international journal that could be broadly construed as the Map of the Mexican Mestizo Genome. The obvious resonance of this name with the Human Genome Project and with the International HapMap Project was shaped by previous publicity. In truth, the first set of results of the Mexican project that INMEGEN officially delivered to Mexico's president, Felipe Calderón, was vastly less grandiose. What President Calderón received was a tentative and preliminary description of the genomic haplotype frequencies (and their variations) among the sampled mestizos as well as an outline of their similarities to and differences from other world populations. Taking some variations as particular ancestry markers allowed researchers to support their much-heralded evaluation of proportions of mestizaje with a likely historical explanation of these under a trihybrid admixture hypothesis (Amerindian, African, and European).[4]

The map made its first public appearance in May 2009, five years after the founding of INMEGEN, when Gerardo Jiménez Sánchez personally handed President Felipe Calderón the fresh-off-the-press scientific paper at the presidential residence of Los Pinos, in front of a contingent of the press. The journal article, together with some peculiar promotional and supporting materials, was packed into a commemorative kit box developed especially for the occasion, colored in a solid colonial blue and adorned with a lively Mexican pink double helix printed on its cover.[5]

A variety of high-ranking government officials attended the formal event, where both INMEGEN's first major achievement and the government's commitment to the improvement of the health and well-being of Mexican citizens were lauded. The ceremony received much press coverage, although some in the media noticed an ironic juxtaposition between, on one side, the unveiling of the map and the explicit positioning of Mexico as a medical research power and, on the other, the state of alert declared after the menacing outbreak of AH1N1 influenza, which was causing a mild panic. Some interpreted the ceremony as an opportunistic maneuver to shore up the public image of Calderón's government as accomplished, effective, and in control of the situation. The timely publication in the *Proceedings of the US National Academy of Sciences* (PNAS) by INMEGEN coincided with the need to give legitimizing publicity to the Mexican government's investments in science. Given the international attention, perhaps INMEGEN's and Calderon's publicists believed that they could manage public confidence in the government's handling of the state of emergency due to the flu epidemic. The occasion was treated accordingly as a highly politicized event.

Calderón, accepting the map, proclaimed that the "deciphering" of the "map of the Mexican genome" constituted a significant step toward advancing scientific research and health in Mexico. He then turned, in the rest of the speech, to the flu epidemic, adding:

> Today when I have the opportunity to be with you at such a singular event, I'd like to thank and congratulate again all Mexicans for their tremendous work of solidarity, prevention, and responsibility in avoiding and reducing the speed of propagation of the new flu virus A/H1N1. . . . In the twenty-first century Mexico faces sanitary challenges that cannot be ignored, and to tackle and overcome them we need scientific research and medicines tailored to the specific needs of every person and every social group. [We need] a more predictive and preventive medicine; this is precisely the advantage offered by genomic medicine. . . . This effort in health will allow us to swiftly and efficiently deal with emergencies like the one we have experienced in the last days. (Sala de Prensa del Gobierno Federal 2009)

Calderón was able to leverage the nationalistic rhetoric in which INMEGEN had framed the MGDP and thus connect the national effort to tackle a flu epidemic with the national effort to develop genomic medicine. The timing of the delivery of the PNAS publication during a sanitary crisis created a cover for INMEGEN against critics who might have questioned the research's lack of immediate concrete medical applications; instead, the MGDP acquired a powerful overtone of urgency and opportunity.

All in all, the press coverage of the launching of the Map of the Mexican Mestizo Genome was very positive, reproducing the sound bites of the day, whose contents were almost identical with those previously recycled through every interview, news report, executive report, and promotional material produced by INMEGEN. The institution's spokespeople repeatedly assured the public that the map was the first major step toward Mexican mestizo genomic medicine. This new medicine was itself nationalistically skewed: it could be fine-tuned to the peculiarities of Mexico's national admixed DNA. It promised not only to be preventive but also predictive and more Mexican. As genomic medical research fine-tuned therapies for the Mexican population, the Mexican state would be enabled to address two of Mexico's greatest public health challenges: the rise in chronic (transitional) diseases such as obesity and diabetes; and the limited funds at the government's disposal to sustain health care. The science was made in Mexico, by Mexicans, for Mexicans, and it was going to put Mexico at the forefront of public health, not only regionally but also globally.

The relative ease with which a recent but powerful scientific institution like INMEGEN could brandish the strongly racialized category of mestizo in the public space and make it the defining element of its gauging of the Mexican population raised few questions for public opinion. Scientists seemed to inhabit the same collective common sense within which most Mexicans found themselves. True, in some scientific quarters, criticism was voiced about the unsuitability of conflating population genomics with a national identity, and the sampling procedures entailed by this assumption (López Beltrán and Vergara Silva 2011).

The authors of the PNAS article demonstrated that the differences between Mexican mestizos from different regional backgrounds in the national territory are caused by differences in the percentage of their ancestral European and Amerindian contributions, that is, by differences in their patterns of mestizaje (Silva-Zolezzi et al. 2009). It is clear that the authors of this article were laying claim to a genomic map of the Mexicans as a necessary reference (and an obligatory passage point) for all future genomic research works related to Mexican nationals, both mestizo and indio. But, given the rapidity with which similar results have been accumulating since (some with higher genomic density) on samples of Mexican and Mexican-Latino mestizos elsewhere, it becomes necessary to evaluate these results in an international context. The desire to present Mexican biomedical research as a national resource that has a monopoly relationship to the deep structural specificities of Mexican mestizos is the effect of a state strategic project. It is still too soon to say if this positioning will succeed in shielding Mexico's so-called genomic sovereignty from constant pressure coming from abroad, or whether this is a prelude to still another stage of INMEGEN's institutional identity (López Beltrán and Vergara Silva 2011; Schwartz-Marín and Silva-Zolezzi 2010).

INMEGEN Changes Gear

At the time of our research, INMEGEN was temporarily located at 1424 Periférico Sur, a bustling speedway not far from the construction site of the edifice that was being built to permanently house INMEGEN, at a cost of 739 million pesos (66 million USD), which were spent on laboratories, offices, conference rooms, a library, an auditorium, and underground parking. When Jiménez Sánchez concluded his first five years as director, in 2009, the building was already three years behind schedule (it was finally finished in November 2012). Construction reports indicated that it had water leakage, structural damage, and defective marble plates, and some stories were circulating about fraudulent expenditure within the construction budget. The unfinished building stood as a symbol of the unfulfilled scientific promises made by Jiménez Sánchez during his tenure at INMEGEN. A blogger appeared who labeled INMEGEN a white elephant and who seemed to have inside knowledge about the huge semiabandoned facility, requiring much care and expense and yielding little profit. Mexico's ministry of health also became dissatisfied with INMEGEN's administration, and after an inquiry that cast doubt on Jiménez Sánchez, he was replaced by Xavier Soberón as director general. The first epoch of INMEGEN thus came to an abrupt end.

Since the arrival of Xavier Soberón in November 2010, INMEGEN has undergone several organizational changes, the most conspicuous of which is a reduced level of public visibility.[6] Important changes in the research personnel ensued, and the staff was shaken up. When former critics of INMEGEN were invited to speak at an important event, a researcher called it "the colloquium of reconciliation." Xavier Soberón had already proved himself as a professor of biomedicine and prominent director of UNAM's Institute of Biotechnology. He used his connections to recruit a number of university researchers from diverse fields (anthropology, medicine, biology) as scientific advisors. Most notably, INMEGEN no longer prioritized becoming a player in the international genomic field. Presumably, this had to do with methodological questions and changes in the international genetic research scene.

Over the past few years, genomics research has to a certain degree experienced diagnostic success: it has identified a number of associations between specific genetic variants and complex human disease (Hardy and Singleton 2009; Hindorff et al. 2009). While genome-wide screening is now touted as the most effective way to identify genes involved in complex diseases, critics point out that single nucleotide polymorphisms (SNPs) discovered by genome-wide association studies account for only a small fraction of the genetic variation of complex traits in human populations (Yang et al. 2010). In addition, the common-variant/common-disease hypothesis that informed the MGDP is controversial and is being reexamined in some circles (McClellan and King 2010). Criticisms relate to the scope and limits of these types of investigations.

Worryingly for the big-science advocates of MGDP, there is increasing evidence that rare variants may play a more significant role in complex disease than do common variants (Manolio et al. 2009; Nelson et al. 2012; Tennessen et al. 2012).

Some of Xavier Soberón's closest advisors advocate whole genome sequencing and the search for rare variants, or for SNPs that might be present in low frequencies in the HapMap reference populations used by the first-period MGDP (but found in high frequencies in Mexico's indigenous population) to be a better approach for estimating admixture and advancing genomic medicine. This, at least, is how a new institutional project of sequencing the first indigenous ("Native Mexican") genome is being justified.[7] As pointed out by a current collaborator of INMEGEN, "identification of variants using genome-wide association studies will have little impact on the Mexican population; the information obtained by whole genome sequencing, and the identification of rare variants within the indigenous population will be of great relevance for genomic medicine and nutrigenomics in Mexico" (Canizales-Quinteros 2011).

Apart from this shift of methodological stance within INMEGEN, few researchers have expressed interest in studying linkage disequilibrium patterns or in doing association studies with the tools derived from the MGDP (even when they are carrying out these types of analysis in the Mexican population). The first scientific article characterizing functional variants of a gene involved in the metabolism of common drugs that uses data from the MGDP's mestizo populations was only published in 2011 by MGDP researchers (Contreras et al. 2011). Another drug susceptibility investigation is being undertaken by MGDP researchers, past and current employees of INMEGEN, which uses the Mexican mestizo panel to explore the frequency of an allele that confers adverse reactions to an antiretroviral drug commonly used by HIV/AIDS patients (Sánchez-Giron et al. 2011). The value of this research, and whether it has an exclusively Mexican scope, remains to be seen.

When our team inquired about the extent to which the MGDP database (publicly available online) had been accessed and presumably applied, INMEGEN's former head of outreach and education commented that in two years it had been downloaded by eighty institutions, only five of which were national. He estimated that 90 to 95 percent of the downloads had come from research institutions outside Mexico and Latin America, mostly in Europe, Asia, India, and the United States. This behavior suggests that the means of communication between INMEGEN and research centers in other countries is by way of databases, directly as a function of the data produced and exchanged.[8]

Communication between INMEGEN and Latin American institutions is routed through other channels; for example, the Peruvian government solicited via diplomatic conduit support from INMEGEN in designing a research project to assess the genomic diversity of their population. Relationships between INMEGEN and other Mexican institutions take different forms, from strategic alliances established horizontally between national institutes of health or between INMEGEN and research centers at various public universities, or one-on-one collaborations between specific research groups at different sites. There are also mixed public-private schemes for advancing genomic medicine in Latin American populations at large, like the Slim Initiative in Genomic Medicine, which also involves the Broad Institute of MIT.

"In Mexico," commented a key informant, "the [Mexican Genome Diversity] project stirred some waters—it caused antagonism in many research groups—but the contributions it has made to the world have been widely acknowledged." If the Map of the Mexican Mestizo Genome (the precious product of MGDP) deserves the hoopla that witnessed its creation, why hasn't it been more widely utilized? The lack of pickup bothers others who have worked on the project since the beginning. In interviews and group discussions, they expressed puzzlement over the research community's lack of engagement with it. When reflecting on the ways in which the MGDP is currently perceived within INMEGEN, one laboratory technician stated, "We have a lot of information on the server that is not being used. It is a pity that people abroad are more interested in using it than are Mexican scientists. In the end, the main objective of the project—that we would all use it and find clinical applications—seems lost."

Speaking of the implications of sharing information on publicly accessible databases, another informant said, "We are publishing data and those who are reaping the benefits are international groups. . . . Before serving the world this information should serve Mexico. . . . If we were a private university this would not be an issue, but we are a national institute of health and we are using federal funding." He later added, "Recent cooperation between INMEGEN and U.S. universities responds in part to wanting both groups to benefit; if the U.S. is going to benefit from our data and from the interesting questions arising from the knowledge of that data, it is better that we both benefit from collaboration." The new approach of establishing joint research ventures with the United States contrasts strongly with the nationalist rhetoric used by Jiménez Sánchez in the early days of INMEGEN, when he discouraged international collaboration yet sought international exposure and spoke constantly of positioning Mexico "at the forefront of medical genomics" in order to forestall future U.S. pharmacogenomic colonization.

Taken together, these facts and the in-house impressions of researchers around the project raise questions about the current use value of MGDP results within the Mexican scientific community. Did the nationalistic character of the MGDP as it was originally conceived and publicized hurt its reputation, in the end, after it was published? Is there any scientific framework under which a national genomics makes sense, or is that concept fatally burdened by an untenable hybridization of politics and biomedicine?

As the MGDP was coming to an end, its main researchers established contact with various multinational collaborative projects, to which they could offer valuable data and expertise in handling that data. It is an interesting comment on the human geography of the international research community that many of these projects are led by prominent Latin American researchers working in the United States and Great Britain (e.g., Carlos Bustamante at Stanford, Andrés Ruiz Linares at University College London), or American researchers of Hispanic origin (e.g., Esteban González Burchard at the University of California, San Francisco). Speaking of the international position of INMEGEN, the former head of outreach and education commented,

> INMEGEN is an interlocutor with genomics institutions in industrialized countries. It is not a follower; it speaks to them face to face. And at the same time INMEGEN . . . is a vehicle for certain initiatives for those Latin American countries that are behind in the development of genomics. When an industrialized country seeks the integration of consortiums in Latin America it contacts INMEGEN first. We give access to informatics tools, academic events. We create synergies. . . . I visualize INMEGEN as a coordinating node that can lend support to other [population genomics] initiatives in Latin America.

What is clear, as INMEGEN turns away from its MGDP focus toward other projects, is that its centralized and distinctively nationalist profile under Jiménez Sánchez has no place within the reticular and large consortium-prone structure of genomics research at large, where the boundaries tend to blur the national affiliations of both institutions and biomedical subjects.

Conclusion

INMEGEN was launched after a campaign orchestrated by Gerardo Jiménez Sánchez and Guillermo Soberón and a small group of influential lobbyists made the establishment of an institution for genomic medicine in Mexico seem like a national priority. The lobbyists succeeded in the short run; funding was obtained from the state for building expensive state-of-the-art technological facilities, and for initiating research on a big-science scale covering the national population. The rhetoric of nationalism under which the institute was launched penetrated the strategic decision to center INMEGEN's energies on a branded population genomic project. Not only was the publicity thickly charged with nationalist and racialist overtones (framed in terms of identity and politics), but so too were the internal parameters of the research.

In particular, the project resonated with the configuration of the mestizo, long an ideal figure in the Mexican imaginary, which encodes the hidden racial identity through which the postrevolutionary state legitimated itself. By positioning INMEGEN as the scientific expression of mestizo pride, the institute swayed public opinion to accept that state expenditure here was going toward something of unique value for Mexicans. The connection of specifically local genetics to the narrative of peculiarly admixed ancestry, and the way this was a unique vector to certain complex diseases, seemed a good rhetorical strategy. However, the clash between the rhetorical image of the science and the research itself provoked distortions and misunderstandings in both public and scientific spheres, exaggerating the reach, the importance, and the immediate usefulness of the project. The scientifically modest MGDP (in its incipient phase) was blown up in the public sphere as the Map of the Mexican Mestizo Genome.

On a purely scientific level, the researchers in charge of the effort were undermined by the excessively high expectations created by INMEGEN's publicity machine. Framed more modestly, the MGDP obtained more than adequate results. A less than perfect but nevertheless very useful sampling and genotyping effort produced an important database, which is now proving to be useful for further research. A good comparative article based on state-of-the-art analysis of haplotype and SNP variation within and between the regional Mexican INMEGEN samples was concluded and published. Several other interesting analyses from that stage are in the pipeline. However, these results ill fit the kind of big-science aura under which the project was announced. The project's publicity has probably had a backlash effect on the actual research.

In the wider social and historical context we have presented, the project reveals the fault lines between geopolitical categories and scientific ones as genomic gauges of population diversity. In this case, we have emphasized the tactically and ideologically crafted notion of the Mexican mestizo that preexisted the project. Scientific research on population genomics was shaped to accommodate the long-standing postrevolutionary effort to forge a bodily and culturally homogeneous notion of the Mexican citizen. The fault lines cropped up because INMEGEN's approach under Jiménez Sánchez was out of sync with the contemporary scientific paradigm of populations. Population genomics, as current INMEGEN research seems to acknowledge, cannot take the geographic and political image of nation-state boundaries and constituencies as authentic scientific parameters. Identity labels like Mexican mestizo are pre-scientific and fluid, and disguise the historical reality of migratory movements and biological interactions.

The use by early INMEGEN scientists of the ideological notion of the Mexican or national mestizo derived from traditional commonsense categories. But the crypto-racial descriptor for Mexicans is a category not well suited for medical population genomics. The effect of circumscribing within national boundaries the Mexican mestizo population put the project at odds with similar ongoing efforts to outline the genomics of Latin American racially admixed groups (including Mexicans) in which the boundaries in the genetic map are drawn ethnically and not nationally. The insistence of INMEGEN researchers on giving weight to a set of SNP variants that were found to be exclusive (until then) to Mexican mestizos and Zapotec Amerindians derived from the political contingencies that defined the sampling, skewing the results for nationalistic, rather than scientific, reasons. Under Xavier Soberón, INMEGEN has changed gear and focus, at the cost of neglecting its former core project. It has adopted a more pragmatic and less nationalist view of population genomics and big-scale genotyping. Admittedly, this change also reflects a global change from the 1990s genomic research paradigm: the international community advancing population genomics has basically turned from genome-wide association studies, to creating thick and complete sets of given genome sequences by fully genotyping particular individuals. Here, a Mexican institution like INMEGEN is well positioned to oversee the full assorting of the genome of an Amerindian subject. This is the route genomic population research at INMEGEN is now taking, but, in contrast to its recent past, the project has not been publicized, and no concrete promises have been made to the public.

Notes

1   A useful anthology that captures some of these works is Bartra (2001).

2   Ruha Benjamin (2009) was one of the first authors to examine the emergence of genomic sovereignty as a postcolonial science policy. She accurately described the paradoxes inherent to INMEGEN's nationalist project, which (re)biologized the nation-state in order to exercise a form of sovereignty that guaranteed scientific autonomy and at the same time invested time and effort to enroll foreign capital and build rapport with global knowledge networks.

3   López Beltrán (2011) uses the term "genomic patient" to describe the person who will—according to the champions of genomic medicine—benefit from the customized health care derived from the individual's genetic makeup. The genomic patient is projected as one who will assume diagnosis, treatment, and intervention as derived from information about her genome. See also Pálsson's (2007) notion of "perpetual patient."

4   The major surprise in the new genomic research is the prominence of genetic markers of African origin in the Mexican populations in most of the national territory, challenging the view that it was concentrated mainly in coastal regions (Cerda-Flores et al. 2002; Green, Derr, and Knight 2000; Lisker, Ramírez, and Babinsky 1996). See chapter 7 (this volume) for examples of how dominant representations of the nation as divided in regions are often racially and ethnically coded. In Mexico, finding markers of African origin throughout the territory was surprising insofar as it clashed with the racialized geography in use, in which the "more European" population tends to occupy the north and the most indigenous population tends to occupy the south, while the coastal regions of the Pacific and the Gulf of Mexico are the only enclaves of African ancestry.

5   The design and content of such a box deserve their own detailed semiotic analysis. Besides a graphic historical timeline that situated INMEGEN's achievement on a par with Mendel's peas, Watson and Crick's helix, and the like, a Spanish version of the article (reproducing the PNAS format) was surrounded by a host of strategically placed announcements in international publications that Jiménez Sánchez used to build up an international presence, most of them more political in content than scienific.

6   Xavier Soberón is unrelated to the INMEGEN lobbyist Guillermo Soberón.

7   According to one of our informants, however, this institutional project was originally conceived under the directorship of Jiménez-Sánchez. "It gained momentum" and is now economically and technologically feasible at INMEGEN, so "the new director chose to continue with it." Another key informant noted, "the indigenous resequencing project is a natural next step from GWAS [genome-wide association studies]." Despite this claim of continuity, INMEGEN scientists in informal conversations sometimes refer to Soberón's preference for dealing with simple, prokaryotic genomes; he is often quoted as saying, "I hate haplotypes," when referring to "complex" human genomics.

8   See chapter 6 (this volume) on epistemic continuity between samples and data, which raises some issues regarding the politics of exchange between INMEGEN and other—especially U.S.-based—institutions.

part II
laboratory case studies

chapter 4

"The Charrua Are Alive"

The Genetic Resurrection of an Extinct Indigenous
Population in Southern Brazil

Michael Kent and Ricardo Ventura Santos

"The Charrua are alive," said the dramatic page-wide headline of an article that appeared in August 2003 in the newspaper *Zero Hora*, published in the south Brazilian state of Rio Grande do Sul (figure 4.1).[1] The news was remarkable, as the Charrua—the dominant indigenous group in the Pampa regions of this state and neighboring Uruguay in precolonial times—were believed to have been extinct since the first half of the nineteenth century. The article was illustrated with drawings of a man dressed in animal hides, brandishing a spear, and with a challenging look in his eyes, and of another man riding a horse sideways, a trademark of Charrua warriors. It did not, however, refer to the small community living in the outskirts of the state capital Porto Alegre that claimed to be Charrua. Appealing to a differentiated ethnic identity, this community had been lobbying the Brazilian state for several years to be granted their own territory. Rather, the article referred to genetic research coordinated by professor Maria Cátira Bortolini of the Universidade Federal do Rio Grande do Sul (UFRGS), located in Porto Alegre. Its opening sentence affirmed that "genes of a people extinct almost two centuries ago may still be present in many Gaúchos—and without their knowing."

"Gaúcho" is the term generally referring to the population of the state of Rio Grande do Sul. In its more specific use, it also refers to the social archetype of the Pampa cowboy that has come to represent the state's distinct regional identity. Since 2001, Bortolini, her postgraduate student Andrea Marrero, and other colleagues have conducted research on the genetic profile of the Gaúcho population. These geneticists placed particular emphasis in their work on the possible Charrua origin of maternal lineages they found among the contemporary population of the Pampa region. Having entered into frequent conflict with colonial populations, the Charrua were defeated by the Uruguayan army in 1831 and most men aged twelve years or older were killed. Four survivors—including their last chief, Vaimacá Perú—were taken to Paris and exhibited as an exotic curiosity, only to die shortly afterward (Bracco 2004; Houot 2002).

Figure 4.1. "The Charrua Are Alive," *Zero Hora*, 18 August 2003 (by permission).

"However," as the article in *Zero Hora* concludes, "according to the work of the UFRGS, the Charrua and Minuano are still alive within many Gaúchos."[2]

It took, however, several more years of research and of careful construction of the argument before the geneticists felt sufficiently secure to make similar claims within a scientific publication. In the third and final publication relating to this research, they assert that "the Charrua maternal inheritance may have been more important than initially suggested" in the contemporary Gaúcho, revealing "an extraordinary genetic continuity at the mtDNA level" (Marrero, Bravi, et al. 2007: 169). The possibility of genetic continuity between the extinct Charrua and the contemporary Gaúchos was already present as a key hypothesis at the start of this research. Over the course of the research process, this idea has taken on different incarnations, within varying contexts, and has been affirmed with differing levels of certainty.

In this particular case, the establishment of genetic continuity was made significantly more difficult by the assumed early extinction of the Charrua. As Bortolini explained during an interview, "I do not have the parameters; I will never know what the Charrua were like because they do not exist anymore. . . . I do not have the original population to compare with."[3] As arriving directly at the genetic characteristics of the Charrua was beyond their reach, the geneticists had to construct continuity with the Gaúchos through what Bortolini defined as "indirect inferences": a multiplicity of associations, exclusions, and translations that established such continuity as probable rather than as definitive. In doing so, she made extensive use of repertoires outside of biology, such as history, archaeology, and social understandings of Gaúcho identity.

In human population genetics in general, inferences and comparisons are a central element of everyday research practice. Affirmations of the identity of specific individuals and populations are always established in comparison with other populations (Barnes and Dupré 2008; Cavalli-Sforza, Menozzi, and Piazza 1994; Pritchard, Stephens, and Donnelly 2000). Anthropological, archaeological, historical, and linguistic data—as well as commonsense knowledge—frequently inform the specific direction of genetic research, as well as the interpretation of findings. The articulation of social and biological repertoires, therefore, is not an exception in genetic research on human populations (Kent 2013; Montoya 2007; Palmié 2007; Santos and Maio 2004; Wade 2007a). Of course, as is discussed in detail in the introduction to this book, such articulations are characteristic not only of genetics, but of the life and natural sciences in general (see Jasanoff 2004b; Latour 1993; Pálsson 2009; Rabinow 1996).

The research on the genetic profile of the Gaúchos was rooted in a long tradition of similar genetic research on human populations in Brazil, which is discussed later in more detail (see also chapter 2; Santos and Maio 2004). In many aspects, the project and the practices analyzed in this chapter can be considered representative of genetic research on the ancestry of contemporary populations in this country. However, they also raise a number of central concerns of this book in a particularly condensed way. Given the assumed extinct status of the Charrua, which made it impossible to arrive at a genetic parameter through the study of a living population, the use of understandings from outside of the genetic field became even more important than usual. As such, this case offers an opportunity for exploring the multiple translations and chains of articulations (Latour 1993, 2005) between social and biological repertoires throughout the different stages of the research process.

The central objective of this chapter, thus, is to gain an understanding of the social and genetic conditions of possibility that have enabled the genetic recovery—one could even say resurrection, given the title of the newspaper article—of the Charrua. It will do so by analyzing the social life of the project, reconstructing the research process from initial hypothesis to the scientific conclusion that genetic continuity between the Charrua and the Gaúchos is highly probable, and through to the social incarnation of this conclusion: the Charrua are alive. More generally, the aim of this chapter is to contribute to a deeper understanding of the multiple articulations, translations, and tensions that occur between social and genetic repertoires across the different stages of a research project's existence.

The Social Construction of Gaúcho Identity

Rio Grande do Sul is the most southern state of Brazil, bordering Uruguay to the south and Argentina to the west. The main indigenous groups occupying this area at the time of colonization were the Guaraní and the Charrua, as well as other hunter-gatherers generally grouped under the label Charrua. During colonial times, control over this area alternated between Portuguese and Spanish hands. After Brazil became independent from Portugal, Rio Grande do Sul waged a war of independence against Brazil's central government from 1835 to 1845, known as the Revolução Farroupilha. A legacy of these historical developments is a strong sense of being different from—albeit part of—Brazil among the state's contemporary population. The construction of a differenti-ated regional identity in Rio Grande do Sul has developed in significant mea-sure as a response to efforts toward centralization by the federal government and in contrast to Brazilian national identity (Oliven 2006).

As mentioned, the term "Gaúcho" refers both to the population of Rio Grande do Sul as a whole and to a particular social type originating in the Pampa grasslands bordering Uruguay and Argentina. The Pampa region, with its extensive cattle ranches, was the center of the state's economic power before the expansion of industry and the acceleration of urbanization in the twentieth century. The Gaúcho archetype is most typically represented as the original horse-riding cowboy of the cattle ranch, wearing *bombacha* trousers, a poncho and hat, drinking *mate*, and eating roast meat. He is also associated with fierce masculinity and a strong sense of independence. As this image offers a clear contrast with Brazilian national identity, the Gaúcho of the Pampa gradually moved center stage in the social construction of regional identity by the state's elites, to the point of nowadays serving as a proxy for the inhabitant of Rio Grande do Sul as a whole. In the process, the image of the Gaúcho became increasingly mythical, taking on iconic proportions (Bornholdt 2010; Oliven 2006). Gaúcho identity is nowadays celebrated mainly through the Movimento Tradicionalista Gaúcho. Created in 1966, it prides itself in being the "largest popular culture movement in the Western world," with approximately 1,500 affiliated centers and over 1.4 million registered members (Oliven 2006: 122–123).

The nineteenth and early twentieth centuries witnessed large-scale migration of Europeans—in particular Germans and Italians—to the state of Rio Grande do Sul. In contemporary constructions of Gaúcho identity, whiteness and European descent play a central role. This claim of being predominantly European is one of the main grounds on which inhabitants of Rio Grande do Sul differentiate themselves from the rest of the Brazilian population. This equivalence between the state's population, whiteness, and European ancestry is also strongly present in genetic research conducted at the UFRGS, particularly in medical genetic studies that use white inhabitants of the state as proxies for European populations (see, for example, Zembrzuski, Callegari-Jacques, and Hutz 2006). In the social construction of Gaúcho identity, the contributions of both indigenous and black people to the formation of the population of Rio Grande do Sul are mostly neglected (Leite 1996; Oliven 2006). The possibility that biological mixture with indigenous people has occurred is an issue that receives very limited attention, including in a significant part of the historiography on Rio Grande do Sul. In this respect, the title of one publication is revealing: "Miscegenation That Never Took Place" (Dacanal 1980). Similarly, until recent decades there have been much more important social barriers to racial mixture in this state in comparison to other regions of Brazil, as evidenced for example by the existence of different social clubs for white and black people (Oliven 2006). As such, the idea of *mestiçagem*—while being a central element in the construction of Brazilian national identity—has only played a minor role in conceptions of regional identity in Rio Grande do Sul.

In spite of the limited emphasis on physical mixture, indigenous populations—and in particular the Charrua—are credited with cultural influences in the formation of the Gaúcho of the Pampa region. The latter are often conceptualized as a cultural fusion of the colonial European and the Charrua, adopting cultural attributes, particular skills, and certain psychological characteristics from both sides. In particular the Charrua's skills in horse riding, their sense of independence, and their bellicose character—some of the most emblematic elements of Gaúcho identity—allow them to be associated with the figure of the Gaúcho (Becker 2002; Oliven 2006). Thus, the indigenous elements that are incorporated into constructions of Gaúcho identity are historically distant in character. References are made exclusively to supposedly extinct populations such as the Charrua, rather than the Guaraní and Kaingang, which have survived into the present and mostly live in situations of extreme poverty. This resonates with a more general tendency in Brazil and Latin America to glorify extinct indigenous populations while at the same time marginalizing contemporary ones (De la Cadena 2000; Monteiro 1996). A hierarchy of value is at play here: from this perspective, the fierceness, hypermasculinity, and independence attributed to the Charrua—particularly in relation to their refusal to submit to European invaders—turn them into more dignified and desirable ancestors than the Guaraní and Kaingang. In the process, the image of the Charrua has been turned into an icon in ways similar to that of the Gaúcho itself. However, it is the former's "psychological permanence" in the Gaúcho "soul" that is allowed for, rather than any physical mixture (Rosa 1957, cited in Oliven 2006: 195).

Origins of the Gaúcho Research Project

The principal investigator of the Gaúcho project, Maria Cátira Bortolini, identifies herself as being of mixed German and Italian descent. With an undergraduate degree in biology, she studied in the postgraduate program of genetics and molecular biology at the UFRGS in Porto Alegre, where she obtained her PhD in 1996. The following year she was appointed as a lecturer in the Department of Genetics. In addition to the research on the Gaúchos, Bortolini has also coordinated research on—among other subjects—the genetic ancestry of black populations, the peopling of the Americas, and the relationship between phenotype and genotype.[4]

The Gaúcho project, started in 2001, was rooted in a long tradition of research on the genetic ancestry of Latin American populations at UFRGS, initiated by Francisco Salzano in the 1950s (Salzano 1971; Salzano and Bortolini 2002). Salzano himself had conducted research on the ancestry of populations of Rio Grande do Sul with classical genetic markers. This included a study on regional genetic diversity within the state, which revealed that the highest proportion of Amerindian ancestry was found in the Pampa region (Dornelles et al. 1999). Bortolini's research, however, was the first to focus on the mythical figure of the Gaúcho of the Pampa. This project also had much in common with other explorations of the genetic profiles of Brazilian populations—both regional and national—conducted at the time (Pena et al. 2000; Santos, Ribeiro dos Santos, et al. 1999). In addition to Bortolini, other key researchers of the Gaúcho project were Andrea Marrero—who obtained her MA (2003) and PhD (2006) degree with this project—and Francisco Mauro Salzano, the latter mainly in an advisory role.[5]

A number of factors stand at the origin of the Gaúcho project. According to Bortolini, one of her primary interests was the possibility of "rescuing" genomes of extinct populations, based on the hypothesis that the Gaúchos of the Pampa might constitute a reservoir of Charrua lineages. As such, the DNA of the Gaúchos represented a typical example of what Bortolini and colleagues elsewhere termed a "witness genome": an admixed genome with the potential to reconstruct the "lost history" of extinct indigenous populations (Bortolini et al. 2004). In addition, Bortolini explained during interviews that she had developed a considerable level of personal involvement with this particular project. The "mythification" of the Charrua was part of her motivation for the research, as she conceptualized the research as a tribute to their last chief, Vaimacá Perú. Bortolini's strong identification as a Gaúcha also played an important part in the choice of research object: "It's the fact that I'm a Gaúcha! A Gaúcho has the motivation to do this. . . . Because I'm proud of being Gaúcha." It was not only the history of the Charrua that the project aimed to reconstruct, but also that of the Gaúcho itself. During an interview in 2003, Bortolini presented the quest for Charrua lineages within the Gaúcho population as "an unparalleled opportunity in the search for knowledge on protagonists of our *own* history" (Salzano and Bortolini 2003; emphasis added). This notion of a quest for knowledge of the self through research on the other is quite familiar in human population genetics. It was, for example, strongly present in genetic research of the 1970s and in the Human Genome Diversity Project of the 1990s, which conceptualized contemporary isolated indigenous populations as proxies for the ancestors of mixed populations or modern societies (for a discussion and critique of this notion, see Cunningham 1998; Reardon 2005; Santos 2002).

Another factor that played a role in the search for the Charrua within the Gaúchos was the geneticists' objective to undermine common associations in genetic research—as well as in popular imagination—between the population of Rio Grande do Sul, whiteness, and European descent. In her research, Bortolini has frequently emphasized the high levels of genetic mixture in Brazilian populations—in particular among people classified as black—with the objective of countering what she perceived as racializing tendencies within Brazilian society (Bortolini 2009; Bortolini et al. 1997, 1999; Hunemeier et al. 2007; Pena and Bortolini 2004). In this respect, the figure of the Gaúcho appears to have been of particular importance. To Bortolini, it represented an overarching social identity capable of overcoming underlying differentiation, in particular in terms of race or skin color. As she mentioned during an interview, "When someone is dressed in *bombacha* trousers, and someone sees him, that person will say, 'Look, there's a Gaúcho,' before saying, 'There's a black person.'"[6]

Academic literature outside of the genetic field, as well as more commonsense understandings of Gaúcho identity, have also significantly shaped the origin and direction of research. According to Bortolini, it was the idea of cultural continuity between the Charrua and the Gaúchos, as well as records of shared cultural elements found in anthropological and historical literature, that triggered the hypothesis of the existence of parallel continuity at the genetic level. The limited emphasis on, and even outright denial of, the possibility of any biological contribution of indigenous people to the constitution of the population of Rio Grande do Sul also contributed to making the exploration of the question of continuity from a genetic perspective relevant. The genetic research on the Gaúchos, then, found itself in constant dialogue with social, historical, archaeological, and linguistic repertoires for defining the identity of the Gaúcho population.

As discussed later, the Gaúcho research constructed the genetic continuity between the Charrua and the contemporary Gaúchos through a number of conceptual steps. First, the geneticists differentiated the Gaúchos of the Pampa from the rest of the population of Rio Grande do Sul, which was realized through an emphasis on their predominantly indigenous maternal heritage; second, they excluded the possibility that the Amerindian heritage present within the Gaúchos could be of Guaraní or Kaingang origin; finally, the third step established connections between the Gaúchos and the Charrua.

The Gaúcho project resulted in three scientific articles that were published in high-impact, international journals. The first and second publications analyzed the genetic profiles of the general population of Rio Grande do Sul (Marrero et al. 2005) and of the Guaraní and Kaingang indigenous populations (Marrero, Silva-Junior, et al. 2007). According to Marrero, these articles "were creating the bibliographic reference base in order to publish the Gaúcho [article]. We were preparing [the way]. The end point was the big article, the article of the Gaúchos with the Charrua." The third publication, which established genetic continuity between the Charrua and the Gaúchos (Marrero, Bravi, et al. 2007), presented in extended detail the objectives, arguments, and main findings of the Gaúcho project.

Research Questions, Methods, and Samples

According to Bortolini, the main question left unanswered by sociohistorical studies was whether in the Pampa region the articulation at a cultural level between indigenous people and colonizers had occurred with or without biological admixture. This is principally a historical question on the nature of social relations between colonizers and the Charrua during colonial times. A number of translations and associations enabled the researchers to address it from a genetic perspective. The formulation of the central question in terms of admixture was crucial, as this is a key area of research in human population genetics. This allowed for its translation into a genetic question on the relative contribution of Europeans and the Charrua to the gene pool of the contemporary Gaúcho.

Another key conceptual step was the establishment of parallels between sociocultural and genetic continuity. Social understandings of the assumed cultural proximity of the Charrua and the Gaúchos—expressed in the form of shared attributes, skills, and character traits—played a central role. Marrero's explanation is particularly illustrative: "The personality of the Charrua is very strong; it is that of the Gaúcho itself. . . . The Gaúcho inherits much from the Charrua: riding horses, drinking *mate*, eating roast meat. So we thought: this Pampean Amerindian component is very strong culturally in the Gaúchos. This must come from somewhere; it can't be something they just absorbed. . . . The cultural inheritance that the Pampa region receives [from the Charrua], I think, is the entrance that allows for the genetic mixture between them." These views suggest that genetic and sociocultural dimensions—including those related to behavior and psychology—of the interethnic encounter are in constant interaction. While on the one hand cultural affinity enables genetic mixture, on the other hand cultural transmission is unlikely without such mixture. This perspective constructs cultural and genetic mixture as indissociable and allows for the translation of cultural associations between the Gaúchos and the Charrua into genetic connections.

The methodology of research consisted in analyzing the mtDNA and Y chromosomes of collected samples, allocating individual lineages to population-specific haplogroups and calculating the relative proportion of Amerindian, European, and African contributions for the sample set as a whole. The choice of nonrecombinant parts of the DNA is particularly appropriate for operationalizing the central research question, as it allows for the establishment of direct connections to the past through a sampled individual's maternal and paternal lineages. This methodology implied that only men could be selected as donors, as women do not possess a Y chromosome.

After defining the central research focus and its methodology, the following step was to build a representative sample set of Gaúchos. The collection of samples focused entirely on the Pampa region, which the geneticists defined as the cradle of the Gaúcho. An additional reason for choosing this region was, according to Bortolini, that "this was the area of the Charrua; this is where you could do this rescue [of Charrua lineages], because there were no other Indians there." Collecting samples in the Pampa region, then, established a first geographic equivalence and historic continuity between the Charrua, the original Gaúchos, and the latter's present-day counterparts.

The first set of samples analyzed by this project—thirty in total—had been collected in the town of Bagé by British geneticist Steven Stuart. These samples had been obtained at the campsite of the yearly celebrations of the Revolução Farroupilha and Gaúcho identity in the month of September (Marrero 2003: 45). The sampled individuals and their parents were all born in the Pampa region. A second set of samples—twenty-two in total—was collected in the town of Alegrete, which was selected because it is located at the heart of the Pampa region. As Marrero explained during an interview, the researchers collected the samples in a central square of Alegrete. Seeing a man with a "Gaúcho appearance," including traditional dress, they would approach him, explain their research, and ask him to contribute a sample. Thus, the social archetype of the Gaúcho, with its specific diacritic signs, partially informed the selection of samples. An additional selection criterion was that sampled individuals were required to have all four grandparents born in the Pampa region.

A third, much larger, set of samples was collected inside the military barracks of Alegrete, which hosted mainly recruits from the Pampa region. As the soldiers had been stripped of possible cultural attributes associated with the Gaúcho archetype through the use of the military uniform, physical appearance became a central criterion in the sampling process. According to Marrero, distinguishing phenotypical characteristics of the Gaúchos such as brown (*moreno*) skin and signs of mixture with an indigenous component became central criteria of selection. A total of 103 soldiers volunteered to participate (Vargas et al. 2006). From each individual, the geneticists collected a blood sample, recorded the soldiers' self-classifications in terms of skin color, and established their own classifications.[7] The place of birth of the volunteers' parents and grandparents were also recorded, in order to ascertain they had an origin in the Pampa region going back at least two generations.

Thus, understandings of the Gaúcho archetype with particular cultural attributes and physical characteristics informed the geneticists' gaze and were encoded into the sampling process. Through the choice of criteria for selecting participants, this process was partly geared toward establishing an association between the social archetype and the genetic Gaúcho. In addition, selecting individuals whose grandparents were all born in the Pampa region constructed the Gaúcho population as immobile and fixed in geographical space. This contributed to the corroboration of the hypothesis of genetic continuity in the region's population between the times before the Charrua's extinction and the present.

The Genetic Differentiation of the Gaúchos

The first conceptual step in the construction of genetic continuity between the Charrua and the Gaúchos differentiated the latter from the population of Rio Grande do Sul as a whole. At the same time it also affirmed the Gaúchos' proximity to the Pampa region of Uruguay. It did so in particular by bringing into evidence a significant proportion of maternal indigenous heritage among the Gaúchos, as well as by constructing their parental heritage as predominantly Spanish.

This step was important for several reasons. In social understandings and historical studies, the Charrua are generally associated with areas under Spanish colonial influence. In contrast, dominant views on the formation of the Brazilian and Gaúcho population leave little conceptual room for inclusion of the Charrua. Thus, in order to construct their argument, the geneticists needed to differentiate the Gaúchos of the Pampa from both the generic Brazilian and the population of Rio Grande do Sul at large. As Marrero explains, "the first thing we did was to separate: culturally and genetically there exist two Gaúchos, which are completely different."

With this objective, the research established the genetic characteristics of samples from different regions of Rio Grande do Sul—used as a proxy for the state's generic white population, labeled as "General RS White" by the geneticists—as well as a sample set from a community of descendants of nineteenth- and early twentieth-century European immigrants in the Serra region, labeled as "Veranópolis" after the community's name (Marrero et al. 2005).[8]

The objective of the analyses was to evaluate the heterogeneity of the ancestry of the population of Rio Grande do Sul (Marrero et al. 2005: 497). Constructing internal difference within this population was a necessary step toward differentiating the Gaúchos of the Pampa. In addition, research results also explicitly differentiated the population of Rio Grande do Sul from the population of Brazil as a whole, by emphasizing the almost entirely European genomic profile of the Veranópolis sample, as well as the significant Amerindian mtDNA proportion in the general sample for Rio Grande do Sul (36 percent) (Marrero et al. 2005), which contrasts with the 22 percent of Amerindian mtDNA ancestry found in a sample of the southern macroregion of Brazil (Alves-Silva et al. 2000). In the third publication of the project, these samples were compared to those collected in the Pampa region (Marrero, Bravi, et al. 2007). The results of mtDNA analysis played a central role in constructing the genetic distinction of the Gaúchos of the Pampa. They showed a proportion of 52 percent of Amerindian mtDNA, leading to the conclusion that they "have the most important reservoir of Amerindian mtDNA lineages in Brazil outside the Amazonian region" (Marrero, Bravi, et al. 2007: 168).

The difference in selection criteria used in the collection of the compared samples has significant effects on the outcome of the comparison between the samples collected in the Pampa region and those from other regions of Rio Grande do Sul. In the case of the Pampa samples, the recorded data on phenotypical appearance reveal a high proportion of individuals that identify as *mestiço* (31 percent), *negro* (6 percent), or *índio* (6 percent).[9] In contrast, both the general Rio Grande do Sul samples and the Veranópolis samples consist only of individuals that were phenotypically classified as *branco* (white) by the researchers who collected and later donated them for the Gaúcho project. Including a significant proportion of mixed individuals in contrast to only white ones and descendants of European immigrants is likely to increase the difference in the proportion of Amerindian mtDNA ancestry found in the respective sets of samples.

Yet there is no mention of these differential selection criteria in the final publication, in which this comparison is made. In addition, in this article the Veranópolis and general Rio Grande do Sul sample sets are pooled together under the label "Rio Grande do Sul other regions," thereby effacing the specific characteristics that are likely to enhance the proportion of European genetic origin. A similar effacement operated in the comparison samples from other macroregions of Brazil, which are also used in the final article without the specification that they correspond to individuals (self-)classified as branco (see Alves-Silva et al. 2000; Carvalho-Silva et al. 2001; and Santos and Maio 2004, for an analysis of similar practices in other genetic research in Brazil). Furthermore, as the number of samples from Veranópolis is approximately four times higher than in the general Rio Grande do Sul set, this pooling of the two dilutes the proportion of Amerindian mtDNA in the set that comes to represent the population of the state as a whole: from 36 percent in the first article to 11 percent in the final publication.[10] The effect is an accentuation of the contrast with the Gaúcho sample set and its majority of Amerindian lineages in the mtDNA.

Analyses using the Y chromosome also played an important role in differentiating the Gaúchos from the population of Rio Grande do Sul. In addition, such analyses associated them with populations more widely recognized as having partial Charrua descent, in particular inhabitants of the Pampa region of the former Spanish colony Uruguay. According to Marrero, the typical Gaúcho of the wider Pampa region is a "mixture of Spanish with Charrua." The geneticists established an association at the genetic level by defining the Gaúchos' paternal heritage as Spanish, rather than the much more common Portuguese ancestry found among the majority of both Brazilians and inhabitants of Rio Grande do Sul. Using analysis of genetic distance and individual haplotypes—among others—they drew the conclusion that "the Gaúcho had four times less differentiation with Spaniards . . . than with Portuguese" (Marrero, Bravi, et al. 2007: 163). This was crucial for the establishment of genetic continuity between the Charrua and the Gaúchos, by embedding the latter within the realm of the assumed generic Pampa mixture of Spaniard and Charrua, and—by extension—within the latter's cultural and genetic sphere of influence.

The Exclusion of the Guaraní and Kaingang

Given the absence of direct genetic parameters for the Charrua, a second key conceptual step in the construction of genetic continuity between the Charrua and the Gaúchos was to exclude the possibility that the encountered Amerindian lineages had an origin traceable to other indigenous groups. The geneticists accomplished this exclusion in two ways: first, by constructing the Pampa region as being under Charrua control at the time of colonization; and second, by excluding the Guaraní and Kaingang—the only two other indigenous populations currently living in Rio Grande do Sul—as possible contributors to the genetic pool of the Gaúchos. This exclusion was accomplished mainly by highlighting differences in mtDNA frequency between the various sample sets.[11]

The publication "Demographic and Evolutionary Trajectories of the Guaraní and Kaingang Natives of Brazil" (Marrero, Silva-Junior, et al. 2007) played an important role in this by presenting the frequencies at which the major Amerindian haplogroups are found in the analyzed samples for both mtDNA and Y chromosome DNA in these populations. In the final article, these frequencies are compared with those of the Gaúcho sample set and with those established for the Guaraní and the Kaingang in the previous article. The difference is particularly significant between the Gaúcho and the Guaraní samples. While the former show an almost even distribution between haplogroups A (30 percent), B (31 percent), and C (30 percent), the latter belong almost exclusively to haplogroup A (85 percent). This led the authors to conclude, "when we considered the mtDNA Amerindian portion, no connection was detected between the Gaúchos and the Guaraní" (Marrero, Bravi, et al. 2007: 169).[12]

In addition, in order to exclude indigenous contributions other than Charrua to the genetic profile of the Gaúchos, the geneticists constructed the Pampa area as being unambiguously under the control of the Charrua at the time of colonization. With the use of several maps based on archaeological and historical data, the geneticists constructed the territorial occupation of Rio Grande do Sul by the Charrua, Guaraní, and Kaingang prior to colonization as mutually exclusive, concluding that there was "absence or little overlap in their geographic distributions since they were traditional enemies and culturally very different" (Marrero, Bravi, et al. 2007: 161). Also, one of the maps defines the area of Charrua occupation as "Pampa Region." Combined with the association made earlier between the Gaúchos and this region, this effectively constructs a continuity in geographical occupation between the precolonial Charrua and the contemporary Gaúchos. These arguments contribute to reducing ambiguity about the potential Amerindian ancestors of the Gaúchos, by backgrounding possible overlaps, interactions, and mixture between the Charrua, Guaraní, and Kaingang. Furthermore, the Kaingang were excluded as a possible source of the Gaúchos' indigenous genetic material with reference to historical records and the analysis of migratory routes, which according to the geneticists showed that the Kaingang had not inhabited the Pampa region (Marrero, Silva-Junior, et al. 2007).

The geneticists also offered a social explanation of why the Charrua were more likely to have mixed with the colonial population than the Guaraní. This explanation had an important gender dimension, with a particular focus on the 5 percent of Amerindian lineages in the Y chromosomes of the Gaúchos, which represents the highest proportion of Amerindian Y chromosome lineages encountered in mixed Brazilian populations to date, surpassing even the Amazon region (Marrero, Bravi, et al. 2007: 163). To the geneticists, this suggested that Charrua men had been significantly more successful at reproducing themselves with colonial European women than men of other indigenous groups, being the only ones to have left their genetic imprint on contemporary Brazilian populations. According to Marrero, it was precisely the Charrua man's resistance and his warrior character—which distinguished him from other indigenous populations—that made mixture with colonial populations possible: it turned him into an object of desire for European women and into a figure commanding respect from men. This explanation resonates well with the mythification of the Charrua discussed earlier. In addition, it offers a different scenario from the general pattern of mixture in the Americas, in which European men were able to impose themselves genetically upon Amerindian and African women through their overwhelming political and economic power, effacing Amerindian and African male genetic material in the process (Pena et al. 2009; Santos, Rodrigues, et al. 1999). In contrast, through this strong emphasis on the Amerindian proportion of Y chromosome data, mixture in the Pampa region emerges as the result of seduction and consent.

Connections between the Gaúcho and the Charrua

Frequencies of mtDNA haplogroups were used not only to differentiate the Gaúchos, but also to establish positive identifications with sets of samples and corresponding populations that are more easily associated with the Charrua. In the first place, the geneticists made connections with the population of Uruguay. Earlier research by Uruguayan geneticist Mónica Sans and colleagues had revealed a high proportion of Amerindian mtDNA lineages in the contemporary population of that country, which they attributed to Charrua origins (Bonilla et al. 2004). In the third article resulting from the Gaúcho project, the authors make a comparison with a sample from the Uruguayan Pampa, which "was also originally inhabited by Native American tribes, including Charrua" (Marrero, Bravi, et al. 2007: 165). Both the proportion of mtDNA (62 percent) and its distribution among the four main Amerindian haplogroups (21/34/32/13) of this sample show relative similarities with the Gaúcho sample set. The same applies to the haplogroup distribution of other reported populations from Uruguay (Marrero, Bravi, et al. 2007).

Second, the Amerindian component of the Gaúcho sample set is associated with indigenous populations of the Southern Cone. The final article's introduction states that archaeological data suggest a "connection between the Charrua and the aborigines of Tierra del Fuego and Patagonia" (Marrero, Bravi, et al. 2007: 161). This article lists the mtDNA haplogroup frequencies of three Patagonian and four Fuegian indigenous groups, based on the extraction of ancient DNA. In a graph of genetic distance based on mtDNA frequencies, the Gaúchos cluster together with these populations at close distance, leading the authors to deduce that there exists "relative mtDNA affinity" between them (Marrero, Bravi, et al. 2007: 169). Similarity consists mainly in a high presence of haplogroup C in the samples from the Southern Cone. Thus, through a series of connections in time and space, the Gaúchos are genetically associated with the probable ancestors of the Charrua and—in extension—to the Charrua themselves. Associating the Gaúchos with indigenous populations that have reached Rio Grande do Sul from the south also serves to further differentiate them from the Guaraní and Kaingang that have reached the state from the north.

In the third place, the article connects the Gaúchos with "the Charrua legendary chief" Vaimacá Perú. Captured by the Uruguayan army during the defeat of the Charrua in 1831, he was taken with three other Charrua to Paris to be exhibited as an exotic curiosity. He died there shortly afterward and his remains were conserved at the Musée de l'Homme. In 2002, his body was repatriated to Uruguay and buried in the National Pantheon. Before his burial, Uruguayan geneticist Mónica Sans performed analyses of his DNA. They revealed that the chief's mtDNA corresponded to haplogroup C (Sans et al. 2010). In their analysis of the Gaúcho samples, the geneticists found no direct matches with Vaimacá's specific lineage. However, because of the importance of haplogroup C in differentiating the Gaúchos from the Guaraní, the presence of this main haplogroup among the only available sample considered as Charrua contributed to the establishment of a genetic connection between the Gaúchos and the Charrua.

The Micropolitics of Scientific Interpretation

The evidence discussed here does not identify the Amerindian lineages found among the Gaúcho samples directly as Charrua. Within the results section of the final article, no association is established between Gaúchos and Charrua. Yet, taken together, these indirect inferences do allow the geneticists to construct this association as probable. In the discussion section of the article, they state, "our results indicate that the Charrua maternal inheritance may have been more important than initially suggested" (Marrero, Bravi, et al. 2007: 169). The language used here is cautious. As Marrero explains, "One thing we have always been very cautious about is that we will never be able to affirm that these lineages are Charrua. And for a geneticist this is very difficult to deal with—you will never be able to settle an issue. But we had near certainty that there was Charrua there. . . . I think that if we would have raised these hypotheses [that the lineages are Charrua], it would have been difficult for us to publish [the article]." However, the article's final conclusion expresses a much more solid level of certainty, claiming, "our data revealed that the known cultural continuity between pre- and post-Columbian Pampa populations was also accompanied by an extraordinary genetic continuity at the mtDNA level" (Marrero, Bravi, et al. 2007). This suggests that different specific contexts, even within one and the same article, allow for expressing a scientific claim with varying degrees of certainty.

Bortolini received frequent invitations to present the Gaúcho project, in particular from universities within the state of Rio Grande do Sul itself. During such presentations she allowed herself more leeway in establishing genetic continuity between the Gaúchos and the Charrua than in scientific publications. In a class taught to postgraduate students at the UFRGS in October 2010, for example, she affirmed, "part of the indigenous lineages that we could not identify as Guaraní, in reality were Charrua; so we succeeded in rescuing a little of the genomes of an extinct people." Informal conversations were another context that offered more room to maneuver. When one of us (Kent) discussed the Gaúcho project with Bortolini and another geneticist, the latter insisted that the evidence pointing toward the Charrua was "*very* indirect." Bortolini, in turn, answered, "but it suggests it, it suggests it. . . . In this case I'm very convinced that this is true. I just can't prove it in a categorical way because of a lack of elements."

Thus, depending on context, affirmations about the connection between the Gaúchos and the Charrua range from total restriction in the data section of academic publications to explicit convictions about its truth in more informal settings, with varying degrees of certainty in between. In constructing continuity between the Charrua and the Gaúchos, the central dilemma the geneticists appeared to face was how to deal with "near certainty" when no technological means were available to offer definitive proof. This placed them within a field of negotiation on the validity of translations between specific research results and more general affirmations about the ancestry of a sample. By making careful calculations about the limits to which it is possible to take the interpretation of data within different scientific contexts—as well as about the amount of probability acceptable to the scientific community—they negotiated their room to maneuver within this contested field.

These micropolitics of scientific interpretation raise a crucial question: why have the geneticists invested such effort in establishing the connection between the Gaúchos and the Charrua? In the first place, this was directly related to a key objective of their research: to recuperate lineages of an assumedly extinct population. Second, defining the indigenous contribution to the genetic makeup of the Gaúchos as Charrua rather than Guaraní resonates with socially acceptable interpretations of regional identity in Rio Grande do Sul. Personal idiosyncrasies also played an important role in practices of interpreting genetic data and constructing generalizations. A brief comparison with Bortolini's research on the genetic ancestry of black populations will illustrate this point.

Bortolini conducted studies on both rural *quilombos*—communities of descendants of fugitive slaves—and urban black populations of several Brazilian towns. A common denominator in this research is a move toward destabilizing genetic continuity between contemporary black populations of Brazil and those of the African continent. Bortolini emphasized the admixed character of the former, in particular through the significant European and Amerindian proportion in their DNA, leading to the conclusion that a "model based mainly on admixture provides a reasonable explanation" for the genetic structure of black Brazilian populations (Bortolini et al. 1999: 557). In this way, her research locates such populations within the contemporary mestiço population of Brazil (see also Bortolini et al. 1997; Hunemeier et al. 2007). Within public debates related to affirmative action policies targeted at black populations, this has led Bortolini to question the validity of identifications of black people and their political movements as being African (Pena and Bortolini 2004).[13] The title she coined for the undergraduate thesis of a student working on black samples from Porto Alegre is particularly illustrative: "Black, but Not That African" (Bisso Machado 2006).[14] However, from a technical perspective at least, the African proportions of 79 percent and 36 percent found respectively in the mtDNA and in the Y chromosome of these samples (Hunemeier et al. 2007) would appear to be more substantial than the evidence that points toward the Charrua in the Gaúcho sample set. Thus, although the genetic data offer considerably more room to maneuver to establish a significant genetic relation between Brazilian black people and Africans than between the Gaúchos and the Charrua, Bortolini has chosen to background continuity in the former case and to actively establish it in the latter.

Such choices need to be examined closely and placed within the context of Bortolini's personal views on regional and Brazilian society. In this case, what is of particular importance is the differential role that she attributed to the categories of black and Gaúcho in mediating social relations. To her, the use of racial categories such as "black" as an organizing principle for the formulation of public policies entailed the risk of increasing racial tensions in Brazil rather than diminishing them. Bortolini proposed instead a focus on the mixed character of the Brazilian population as a means to combat racism (see chapter 7). In contrast, for her the importance of the social category of the Gaúcho rested in its unifying ability to transcend underlying racial and social differences. Revealing the incorporation of a high proportion of Amerindian genetic material in the contemporary Gaúchos—in particular if it could be traced to supposedly extinct populations—was an important element in the construction of the Gaúchos as more inclusive at the genetic level as well. It also served to destabilize common, racialized identifications of the population of Rio Grande do Sul as white and European. At the same time, it upheld the distinction of the Gaúchos from the general Brazilian population, but grounded it in a different kind of mixture rather than in an argument of European exclusivity.

The Charrua Are Alive: Social Engagement with the Gaúcho Project

The question of interpretation raises different issues when genetic knowledge moves beyond the scientific field and is engaged by, for example, the media, social movements, and others. When such knowledge is incorporated into public debates about ethnic, racial, national, or regional identity, it starts to interact with a variety of preexisting discourses for the definition of individual and collective identities. Social interpretations of genetic data often reveal a predisposition toward establishing an alignment between such data and collective identity constructions, political interests, or personal trajectories. A frequent effect of such practices is the conflation of social identity and genetic ancestry, thereby contributing to the emergence of "imagined genetic communities" (Simpson 2000): collective identifications among a group of people based on shared genetic characteristics (Brodwin 2002, 2005; Gaspar Neto and Santos 2011; Kent 2013; Nelson 2008b; Pálsson 2008; Wade 2007a).

In the case of the article that appeared in *Zero Hora*, it was the journalist that was willing to go well beyond Bortolini in the affirmation of continuity between the Charrua and the Gaúchos. Asserting that the former are "alive" within the latter takes the recovery of mtDNA lineages symbolically within the realm of the resurrection of a supposedly extinct indigenous population. When one of us (Kent) asked Bortolini what she thought about the feature's title, she answered with a sigh of resignation: "Well, that's journalists' stuff. I made all the reservations during the interview, but then they gave it this sensational title." However, she did not object to the title. In continuation, Bortolini said, "But I think it's nice. I liked this one in particular, because in this specific case it's true. . . . If you look at it in a romantic way, they were not extinguished."

Although Bortolini would refrain from making such an affirmation to a scientific audience, it represented a translation of her ideas for circulation in public debates with which she felt comfortable. Thus, the journalist took off from the point where the constraints of the scientific field put definitive limits on Bortolini's own room to maneuver. He operated a translation from the scientific affirmation of likely genetic continuity in the matrilineal line between the Charrua and the Gaúchos into a social language thought to resonate with his audience: that "the Charrua are alive." The imagery of the Charrua to which the article appealed was precisely that of the mythified, fierce warrior with which the public in Rio Grande do Sul can easily identify. This enhanced the social acceptability of the article's claim of continuity of the former within the latter.

This newspaper article led to significant engagement with the results of the Gaúcho project by individuals and social segments beyond the scientific field. Bortolini received a number of requests for genetic tests from people who claimed to be of Charrua descent:[15]

> They came saying, "Professor, I dreamed I was Charrua" or "I want to know whether I'm Charrua, because I have an untameable soul." All things like this. . . . Never, never did anyone come to me claiming to be Guaraní. The Guaraní has no appeal to them. He doesn't have this Hollywoodesque appeal of the Charrua. . . . It's like my husband, when I tested his DNA and I found a lineage you have among the Kaingang. He said, "Did it have to be indigenous people who didn't even use feathers on their heads?"

Marrero reported similar reactions among fellow students at the UFRGS when she told them that genetic evidence pointed toward a much higher proportion of Charrua than Guaraní ancestry: "One person said, 'Well, thankfully, because the Guaraní bowed to the colonizers and the Charrua did not.' So even that is a source of pride. [They said,] 'The Indians from which I descend were fierce, warriors, fighters. They died staring their executioner in the eye.' This was strong [in the reactions]." Thus, for a share of the population of Rio Grande do Sul the extinct Charrua are an object of desire, mainly because of their resistance to colonization and their "Hollywoodesque" appeal. This reveals again a hierarchy of value, in which fierceness, rebellion, and diacritic markers of indigenous distinction such as feathers play a central role. In this hierarchy, the Charrua rank highest.

Positive reactions to the research also came from the Pampa region itself. According to Marrero, many people in this region acknowledge that they are mixed but do not identify with the Guaraní. When the results of the research became known, she received several reactions such as, "I knew it—I knew that this color of mine was Charrua." This reveals an anxiety about having to interpret one's skin color and mixed descent in ways that are socially acceptable. The results of this research offered the Gaúchos an interpretation of their ethnic or racial constitution that was more in line with their own aspirations. Having a darker skin because of Charrua descent was conceptualized as more socially acceptable than having Guaraní—or black—ancestors.

Research results were also picked up by movements of Gaúcho nationalism. They were featured, for example, on a website celebrating regional identity, together with affirmations of pride in being part of the people of Rio Grande do Sul and that the "essence" of the Gaúcho "persists" (cited in Kent and Santos 2012b: 363). The social acceptability of this research for the Gaúcho movement appears to be associated mainly with the possibilities it offers to establish difference with the generic Brazilian population. Finally, the leaders of the small Charrua movement that were campaigning for a territorial concession from the Brazilian state also contacted Bortolini on several occasions. Rather than asking for genetic tests, they requested her to publicly endorse their claims to Charrua identity. They hoped to obtain scientific support for these claims, which were heavily contested by state officials and other players in the political field on the grounds that the Charrua were extinct. After Bortolini repeatedly declined, arguing that science did not provide her with the means to make such assertions, they did not pursue the issue further. However, as the Charrua cacique Acuabé explained to one of us (Kent), although they did not receive the support they had hoped for, Bortolini's research did become instrumental in obtaining official recognition for their claims, as it contributed to the acceptability of the idea that the Charrua had persisted into the present (for a comparable case of the Uros in Peru, see Kent 2013).

Thus, a number of factors stand at the source of the social engagement with—and acceptance of—the Gaúcho project by the public at large. The resonance of research results with social interpretations of Gaúcho identity played an important role. While the geneticists revealed a predisposition to search for the Charrua within the genetic material of the contemporary Gaúchos, the willingness among the general Gaúcho public to accept this association was even more evident. What enabled the translation of genetic results into the idea that the Charrua are alive was in particular the social acceptability of having Charrua ancestors in the Gaúcho social imagination, as well as the possibility this research offered to displace the Guaraní to the background as possible ancestors of the Gaúchos. Other factors contributing to the social acceptability of the Gaúcho research were more distinctly political, as in the case of the Gaúcho nationalist movement and the Charrua movement.

Conclusion

The social life of the Gaúcho project reveals the strong articulations between social and biological repertoires in the everyday practice of genetic research. At all stages of the research process there was a constant interaction between genetic ideas and practices, on the one hand, and archaeological, anthropological, and historical literature—as well as commonsense knowledge on Gaúcho identity—on the other hand. The claims that the geneticists made about the genetic constitution of the Gaúcho population were based on their analysis of DNA material. Yet the specific terms in which they addressed this issue were strongly influenced by preexisting social and historical ideas on the identity of the Gaúchos of the Pampa, as well as the social and symbolic relations this identity condenses. It was the central role of the extinct Charrua in the Gaúcho imaginary that enabled Bortolini to raise the hypothesis of genetic continuity between them in the first place. Over the years, the researchers little by little created the conditions for the establishment of such continuity through a sustained effort at constructing inferences, associations, and exclusions. At the other end of the research trajectory, it was in particular the preference among many Gaúchos for Charrua rather than Guaraní ancestors that made this research socially acceptable to a wider public. It created the conditions for the translation of the genetic conclusion that "male samples of the Pampa region reveal a high proportion of haplogroup C in their mtDNA" into its social incarnation, "the Charrua are alive." This analysis of the trajectory of the Gaúcho research reaffirms arguments made throughout this book about the biosocial or hybrid character of scientific practices and categories, as well as the impossibility of disassociating social and biological domains (Jasanoff 2004b; Latour 1993; Pálsson 2009; Rabinow 1996; Reardon 2005).

In many respects, through this research the geneticists reconfirmed conventional social interpretations of Gaúcho identity. They located the authentic genetic Gaúcho firmly within the Pampa region. They also constructed the Gaúcho as hypermasculine—partly a result of the standard methodology they employed—and contributed to its mythical status by attributing to it heroic origins in a warrior indigenous population. In addition, the geneticists provided scientific support for the Gaúchos' claim of being different from the generic

Brazilian population. Finally, they accorded only limited attention to the role of the black population in the formation of the Gaúchos.

However, in other ways this research also significantly reconfigured the question of Gaúcho identity. In the first place, it achieved the incorporation of genetic repertoires within the public debate waged around the Gaúchos, thereby resulting in a measure of biologization of this identity. The conflation of social identity and genetic ancestry by people outside of the scientific field further contributed to the emergence of an imagined Gaúcho genetic community. In the process, the hybrid character of the Gaúcho itself was accentuated, as its social and genetic characteristics became increasingly associated. The additional conflation in social constructions of regional identity of the Gaúcho of the Pampa with the population of Rio Grande do Sul as a whole means that affirmations about the genetic constitution of the former may well become projected more widely upon the state's entire population.

In addition, this research reinserted a physical indigenous component within constructions of Gaúcho identity, by affirming the occurrence of high levels of mixture between colonizers and indigenous populations against conventional interpretations that downplayed the possibility of such mixture. It converted indigenous people into socially acceptable ancestors, no longer a radical other, but now made familiar through their incorporation into the bodies of the Gaúcho population. However, it was not indigenous populations in general that became incorporated into the imagined genetic community of the Gaúchos, but specifically the Charrua. As in the social construction of Gaúcho identity, it was a historically distant, supposedly extinct and glorified indigenous people—in addition diluted within mestiço bodies—that became part of this community, rather than indigenous populations that have persisted into the present. Given the positive values attributed to the Charrua and the more negative ones associated with present-day Guaraní and Kaingang, in order to gain wider social acceptation this reinsertion almost inevitably had to pass via the Charrua. While Bortolini's explicit objective was to rescue extinct genetic lineages, her personal engagement with the research suggests a concern with rescuing more than that in the process. In the first place, this research takes the Charrua out of oblivion and brings them back into existence within another population. Second, it aims to recuperate what Bortolini sees as the inclusive character of Gaúcho identity, which in everyday practice is consistently undermined by the tendency among the population of Rio Grande do Sul to emphasize European exclusivity and to downplay indigenous and African origins.

For present-day indigenous populations of Rio Grande do Sul, the social consequences of this research are in potential both positive and negative. On the one hand, it increased the social acceptability of having indigenous ancestry for the population of Rio Grande do Sul. In addition, for the contemporary Charrua community, the notion that the Charrua have persisted into the present had an empowering effect in their struggle for official recognition and territorial rights. On the other hand, however, this idea of the genetic continuity of the Charrua may also serve to relativize their extinction and to dilute responsibility for the genocide of indigenous populations in Rio Grande do Sul. And in the case of the Guaraní and Kaingang, their marginalization from contemporary society has now become paralleled by their exclusion from the imagined genetic community of the Gaúcho. In significant measure, the same can be said for the black population of Rio Grande do Sul.

Finally, this research and its social impact reveal a dynamic interaction between the ways in which the Gaúchos and the Charrua are imagined and reflected upon. While the supposedly extinct Charrua are resurrected through the contemporary Gaúchos, the latter are imagined, reaffirmed, and reconfigured through the Charrua. As is common in the construction of collective identities (Anderson 1983; Hobsbawm and Ranger 1983; Sahlins 1999), in this case as well the reconstruction of the past is inextricably connected with the reconfiguration of the present. In the process, the geneticists not only contributed to the continued mythification of Gaúcho identity. They also, as the title of the newspaper article suggests, literally revived the Charrua myth.

Notes

This chapter represents an extended version of an article published in Portuguese in *Horizontes Antropológicos* 37 (2012).

1   We thank the newspaper *Zero Hora* for their kind permission to reproduce the feature "Os charruas vivem."

2   Charrua not only refers to the people of the same name, but is often also used as a generic label that includes other indigenous populations living in this area in precolonial times, such as the Minuano, Guenoa, and Yaró.

3   The data analyzed in this chapter are derived from two complementary ethnographic sources: interviews and observations during fieldwork and analysis of scientific publications. Where data have their source in publications, reference is made to the respective scientific text. Where no specific reference is given, the data come from interviews and observations.

4   See Bortolini's online curriculum vitae for additional information on her research trajectory and publications ("Maria Cátira Bortolini," CNPq, http://buscatextual .cnpq.br/buscatextual/visualizacv.do?metodo=apresentar&id=K4784272U6).

5 For a full list of collaborators of the project, see Marrero et al. (2005), Marrero, Bravi, et al. (2007), Marrero, Silva-Junior, et al. (2007).

6 In order to illustrate this point, during presentations Bortolini showed a photograph at a cattle ranch of several men of varying skin color dressed as Gaúchos.

7 The applied categories were *branco*, *mestiço*, *negro*, and *índio*.

8 These samples had been collected by other geneticists in the frame of medical and forensic genetic research, and contributed to the Gaúcho project.

9 Figures featured in Bortolini's archives of the project.

10 There were 119 from Veranópolis against 31 from Rio Grande do Sul, according to numbers listed in Marrero et al. (2005: 497).

11 As such, this research did not conceptualize the Charrua, Guaraní, and Kaingang as being biologically different in significant ways, but rather pointed toward evidence of different ancestries and histories of migration.

12 Note, however, that in the same article the authors examine specific mtDNA lineages that have offered perfect matches with others found in the literature. These matches can all be directly or remotely traced to the Guaraní (Marrero, Bravi, et al. 2007: 167). These findings, however, have not been given interpretative priority.

13 See the introduction and chapter 2 for a more detailed discussion of these debates and Brazilian systems of racial classification.

14 Negros, mas nem tão Africanos.

15 Bortolini ended up testing the mtDNA of five individuals, all of them revealing European matrilineal descent.

chapter 5

The Travels of Humans, Categories, and Other Genetic Products

A Case Study of the Practice of Population Genetics in Colombia

María Fernanda Olarte Sierra and Adriana Díaz del Castillo H.

> By experience and by affinity, some of us begin not with Pasteur, but with the monster, the outcast.
>
> —Susan Leigh Star, "Power, Technology and the Phenomenology of Convention"

During an eight-month ethnography project, we met a group of young geneticists practicing nonhegemonic and original forms of science, while at the same time searching for ways in which to navigate the mainstream world of population genetics. In this chapter we analyze this group's endeavors regarding the development of population categories, and how such categories refer to a particular ideology of science and scientific practice. To do this, we describe the planning, conducting, and writing up of Project Guajira, a research study aimed at describing one of Colombia's departments, with a special focus on the indigenous group inhabiting it, the Wayúu.[1] We focus on the different population categories produced by these geneticists throughout the research process.

While carrying out our fieldwork we observed that, in the lab where Project Guajira was conducted, the population categories used in the research resulted from a highly dynamic negotiation process in which the researchers' struggles to account for diversity and to reflect on the consequences of their own practice were pervasive. In this process, they produced innovative population categories that differ from those commonly used in genetics research in Colombia, namely mestizo, Amerindian, and Afro-Colombian. Despite this effort, there was a point at which discourses and practices became flattened. The categories not only became discrete and static, they actually returned to the classical forms in order to enable interaction with national and international peers; a first article was written and submitted to a journal with the innovative categories excluded. But then another major change followed. After a process of reflection on the article, the researchers decided not only to withdraw it from the journal but also to write a completely new one. The new article, using the nonstandard population categories developed during the research, represents more closely the research interests, methods, and political stance of the group.

We consider that studying these shifts will allow a more comprehensive view of scientific practice in general, and of counterdiscourses within the field of population genetics in particular. We develop the chapter in four steps. First, we briefly contextualize genetics research in Colombia as it appears in scientific publications. Second, we present the lab that we focus on. Third, we use the example of Project Guajira to illustrate the ways in which these scientists negotiate the categories and methods they use. Finally, we address the movements by which the group's dynamic and manifold categories became static and singular as they traveled to an article submitted to an international journal, and then how later they reappeared in their innovative form.

Colombian Genetics Research on Paper

Published research on Colombian human genetics, especially on population and forensic genetics, is based on two interconnected assumptions about the country and its inhabitants (though some medical and epidemiological research is also based on these same assumptions) (Arias et al. 2006). The first is that the Colombian population is the product of a mix between Europeans (mainly Spanish conquerors), traded slaves from Africa, and the Amerindian indigenous populations. These multiple origins make the population a tri-ethnic one (Rey et al. 2003). In consequence, the categories by which genetics research is organized and published are mestizo, understood mainly as the offspring of European men and indigenous women; Amerindian; and people of African descent (Bedoya et al. 2006; Paredes et al. 2003; Romero et al. 2008; Rondón et al. 2008). This way of understanding the country's population is not limited either to genetics or to Colombia, however. As described in the introduction of this book, the ideology of racial admixture (in Spanish, *mestizaje*) cuts across Latin America. From the late 1800s and early 1900s, and with marked differences from country to country, the mestizo became a token for national unification in political and population terms, and today is considered to represent the main population group (Carrillo 2001; Olarte Sierra 2010; Olarte Sierra and Díaz del Castillo H. 2013; Stepan 1991; Suárez y López-Guaso and Ruíz-Gutierrez 2001; Wade 1993).

This brings us to the second assumption by which Colombian genetics research is organized, performed, and published: that Colombia is divided into a number of cultural and geographical regions, whether five (CINEP 1998) or four (Paredes et al. 2003). Each region is believed to host people with different degrees of admixture (Paredes et al. 2003; Yunis Turbay 2009; Yunis et al. 2005), and it is agreed by many geneticists that due to the perceived difficulty of the country's geography, regional human populations were relatively isolated until improved communications made it more feasible to move from one region to another (Bedoya, personal communication, 2010; compare Carvajal-Carmona et al. 2000). "The population of mixed ancestry concentrates mainly in urban areas, particularly on the Andes. African-Colombians live predominantly on the Caribbean and Pacific coasts and islands. Native American populations concentrate mainly in the East (on the vast Orinoco and Amazon River basins) and in rural areas of the Southwest and North of the country" (Rojas et al. 2010). This understanding can be considered problematic if one unpacks the idea of being mixed. Some of the geneticists we interviewed recognized that in Colombia all populations are, to some extent, mixed. Yet the portrayal of the country in the quotation above mobilizes an idea of the people as geographically organized by racialized categories, which are said to be genetically different from one another.[2]

GPI: Crossing Boundaries

We conducted ethnographic work in four genetics laboratories in Colombia. Here we present data from the lab in which we did the most intensive ethnographic work for over eight months in 2010: the Population Genetics and Identification Group (hereafter GPI, its acronym in Spanish) of the Genetics Institute of the Universidad Nacional de Colombia in Bogotá. As we shall show, we consider this laboratory to be distinctive in the way in which the researchers conduct research and tackle their interests, and in the fact that the head of this group is a young—albeit experienced—geneticist who recently obtained his PhD.

In this context, and following Star (1991) and Law (1991), we chose not to focus on "heroes, big men, important organisations, or major projects" (Law 1991: 12). Our aim was to avoid falling into "managerialism" and to refrain from producing an analysis "filled with active, manipulative agents who stand some chance of ad-hocing their way to organisation and success" (Law 1991: 13). Focusing on "the ordinary folk instead of the big men," as Law (1991: 12) puts it, we provide a view on science that is not about the highly successful, visible, state-of-the-art research groups. By foregrounding those who, for reasons that we address next, can be considered somewhat outside mainstream scientific work, while still clearly belonging to the scientific community, we reveal the multiplicity inherent in scientific practice.

Our interest in focusing on this group relates to its specific nature, which we characterize as marginal. This marginality is not a question of going unrecognized by other geneticists as peers, but rather a matter of the way in which their innovative ideas come to wrestle with and unsettle well-established scientific concepts. In other words, we argue for studying diversity and innovation instead of looking at the usual, already well-known scientific groups. We consider marginality to be a privileged location from which to study tensions within science and thus to be able to advance a plural and nuanced view of human population genetics.[3] In this sense, we do not pretend to depict the group we present here as unique and utterly different from other research groups in the country; rather, we aim to show how research is varied, multiple, contingent, and in many ways contradictory.[4]

The GPI is a rather young research group. It has only been in existence since 2005 and its first population genetics projects started in 2008. Thus its production, in terms of published articles, is rather slim. On the other hand, the group produces many bachelor's and master's theses (with the majority of group members having recently finished their undergraduate studies or just beginning their graduate studies[5]), carries out a number of research projects, and enjoys a growing recognition for the lab's certified and accredited services in forensic genetics. It is important to note that the group's work focuses on population and forensic genetics, the latter in the form of paternity testing. The group's researchers consider it important, however, that all their samples are collected for population genetics purposes and are not derived from either forensic or medical genetics; this allows them to collect additional information to help make sense of and contextualize the results, which is one of the features that makes this group different from the others we studied.[6] Although the GPI is only beginning to make a space of its own within the scientific community, its director, William Usaquén, is widely recognized and respected in genetics networks, both locally and globally. Among other things, he coordinated a group that developed the first attempt to construct a genetic map of Colombia (Barragán Duarte 2007).

The GPI is part of the well-known genetics institute of the Universidad Nacional de Colombia, which belongs to the School of Biology. It is the only group within the institute that works on population genetics, hence our choice to work with them. Their main objective is to characterize the Colombian population as a whole, accounting for the origins of the population and the peopling of the country, and to describe the current genetic configuration of Colombian people. To do this, they combine genomic and genetic findings with demographic, anthropological, archaeological, and ecological information. They are interested in doing more than merely generating genetic data, as Usaquén expressed:

> Innovation in the field of genetics is out of synchronization with our own understanding of it. . . . One can now sequence eight thousand samples in a week. So what? What do all those data mean? If you don't put them in context they mean nothing. It doesn't matter that journals publish articles giving the illusion that genetic data alone are able to convey some sort of meaning. . . . No, if you don't know the history of the place, the culture of the people you are studying, marital practices, commercial networks, and so on, you've got nothing, even if you have all the allele frequencies, SNPs [single nucleotide polymorphisms], mtDNA [mitochondrial DNA], and anything else you want.[7]

In this sense, geneticists at the GPI understand and perform genetics as a science that needs to be articulated with other kinds of knowledge in order to be meaningful. Within this context, the GPI seeks to contribute to advances in methods of data collection and analysis that will enable the incorporation of different types of information that they consider relevant. However, the GPI differs from other labs we studied not only in its understanding and performing of genetics, but also in terms of whom it considers a relevant audience for the dissemination of research findings. The GPI believes that one of the main target groups for disseminating results is the populations they study, as a means of contributing to their understanding of their own history.

Given the GPI's distinctive path of research and commitment, one might wonder about funding. In this regard, it is noteworthy that although the GPI receives funding from the university in which it functions, and the youngest members receive research grants from the Colombian Department of Science, Technology, and Innovation, the earnings derived from paternity testing are becoming a paramount and reliable source of income. Therefore, their research agenda is not necessarily linked with the priorities of funding agencies (which, in Colombia, are mainly related to genetic epidemiology). Earnings from paternity testing thus provide a degree of research freedom that translates into the kind of research that the GPI conducts.

Ways of Becoming

As just shown , the GPI turned out to be different from the other labs we studied, and their distinctiveness has to do specifically with the group's stance on scientific practice and production. This mode of practicing genetics is interiorized, in part, through lectures, discussions, and debates led by Usaquén. Often we heard Usaquén encouraging lab members to be reflexive about their work and the knowledge they gain and produce: "The geneticist is a star in the lab, but if he loses the knack [for using the micropipette] he's lost. He has no idea about the context. . . . The geneticist is impressive because of the white coat and the laser, and the story about being all microscopic and molecular, and because of the idea that he cannot be wrong [because what geneticists do is science]. But he can be wrong; actually many times he is wrong. This [genetic information] is just another datum, nothing more . . . and that's why it needs to be contextualized." Such interventions were intended to put in parentheses the unquestioned accuracy of science and its foundations, and the neutrality of scientists. They enabled discussions in which lab members did not always agree with each other, leading to the further problematization of their own endeavors. In this sense, the researchers working under Usaquén are constantly reminded to be reflexive, to remember that research, practices, results, and interpretations are not neutral. They are called to become visible and "unmodest," which necessarily pulls them away from hegemonic ways of performing science (Haraway 1997).[8] The case presented in this chapter illustrates how this discourse translates into practice, that is, in the development of innovative population categories for human genetics research.

In the move toward becoming visible and unmodest, Usaquén is not only reflexive about his own work and the work of those under him; he also acknowledges that his take on science and the scientific method is influenced by his personal history, his teachers, friends, interests, and trajectory. In short, he situates his knowledge (Haraway 1988). After graduating with a degree in biology, Usaquén worked as a forensic entomologist, took history courses and photography classes, and specialized in statistics. For him, all of these interests have influenced his work on genetics and have provided him with his present viewpoint:

> Many of the things I do as a teacher and as a geneticist are because I learned
> them from Prof. IB. For example, the way in which I relate to you, to my
> students. She taught me the importance of generosity when teaching, to
> teach with patience, to have time to discuss ideas with students. However, I

also have to say that I address research problems differently. The way I think and the way we think as a group isn't the same as when I was her student. . . . The School of Biology here also made me what I am. It was a rather humanistic school; we were encouraged to think critically, to not take things for granted. . . . Another person crucial to my development as a biologist and as a geneticist is my dear friend AC. He's like my brother—you know him. He's a biologist and a philosopher and he makes me question my own assumptions. . . . You have to always remember that personal life defines you as researchers.

Thus, being a scientist is about becoming one. No news here. But we want to highlight that being the kind of scientist that these geneticists are involves being knowledgeable about one's own path and becoming a more reflexive, unmodest, and situated geneticist. Therefore, the GPI as a group is shaped not only by Usaquén's own history, but also by the personal histories of all those working with him. After a presentation about other geneticists' work, Madelyn Rojas, a junior researcher in Usaquén's team, realized their work was different from other research groups. Usaquén answered her question about the reasons for that difference, saying, "This group is not what it is only because of me; you also have and bring your own history. We all give direction to what we do. If you were different I would also be different, and the group would be different." We could say that one requirement for being a GPI member is a willingness to reach this level of reflexivity. It is with this recognition of every member's trajectory, interests, and limitations, and of their shared, nuanced view on human genetics, that the GPI's research projects and agenda are formulated.

Project Guajira: Innovations, Negotiations, and Compromises

Members of GPI consider that their work should contribute to building a positive national identity by celebrating the diversity of the country's peoples, while at the same time keeping alive the history of ancestral groups. With this in mind, the group embarked on a project that aimed to characterize the department of Guajira, recognized as having various ethnic groups. We chose to focus on this specific project for several reasons. First, being conceived as part of the GPI's overall objective of characterizing the Colombian population, it is suitable as an example of the lab's interests and approach. Second, almost all members of the lab participated in the project, whether collecting or analyzing data or writing up the results. Third, it is a good example of the compromises involved in the process of knowledge production, which help us bring to the surface possible tensions in science, while also allowing us to see the input provided by different members of the group. This project and the articles produced from it were a continuous source of discussion, dialogue, and negotiation between group members during the time of our fieldwork and after. Last, this project's results will constitute one of the first published articles of the lab as a research group. We consider it useful to look at the issues at stake when laboratory and communication practices need to be aligned. We show how the long process of negotiating the writing of an article (and the way the researchers positioned themselves within the world of population genetics) translated into a submission to a scientific journal, subsequently withdrawn, after which a rewriting process took place, the outcome of which formed part of Usaquén's doctoral dissertation.

The effort of remaining true to the group's stance and still trying to gain a place in the world of population genetics journals is shown in the following excerpts from our field diaries.

3 June 2010

Today the group discussed the Project Guajira article for two hours. They couldn't agree on how detailed the methodology should be. Usaquén emphasized that the methodology should mention the long fieldwork, the materials used by them, like the poster and the survey. Also, he was emphatic that the article should say that they did not take blood from indigenous people since for them blood is sacred and cannot be given away. But researcher Eljach felt uncomfortable with including that kind of information. He said, "This is a scientific journal and all that social stuff is irrelevant for them. It may even be counterproductive for the paper." In the end, they could not agree on how much to include in the methodology. Usaquén proposed that in two weeks' time each author should bring a proposal on how to write the section. Before leaving, Usaquén said, "Remember that our trademark is the way in which we collect data; if we don't include that, we are ignoring what makes us different from the rest."

17 July 2010

The group discussed the Project Guajira paper. Researcher Sarmiento presented an outline of what she wants the paper to be. Last week they agreed that she should be the first author, as this paper will follow from her BA thesis. She wants an article that combines demographics, genetics, and ecology. Again, researcher Eljach was not convinced. He said, "Is it going to be a demographics article that combines genetics tools or the other way around. . . . I don't see how this will work. I don't think we have enough demographics material [in terms of quantity and statistical representativeness] to make a strong point and we run the risk of ending up with no argument. . . . We should keep it only as a genetics paper."

Researcher Sarmiento replied, "Well, no, we do have enough material to make it a good article. It won't be intuitive; it will be a robust argument. And if we don't present the demographics we won't be able to account for the new categories we are proposing."

6 October 2010
   The group is still debating whether or not to produce one paper that combines demographics with genetics. The group is very much divided on this issue. For Dr. Usaquén, and researchers Sarmiento and Alonso, the data are sufficient for making their argument strong. However, researchers Casas, Eljach, and Rojas insist on the weakness of their demographic data. Also the latter three fear that including too much nongenetic stuff will harm the paper and will diminish the possibilities of getting published.

The overall discussion revolved around how, for some researchers, the inclusion of so-called social data would make the article unsuitable for publication, while for others the articulation of this kind of data with genetic data would be the group's main contribution to genetics. The negotiation over what kind of data to include shows, on the one hand, that there was no easy or single decision on how to make their research public. On the other hand, it shows that within the group different takes existed on what their scientific publications should look like, in terms of the kind of data deemed (ir)relevant for mainstream genetics. We interpret the negotiation as an example of how this group of scientists is reflexive about the kind of geneticists they want to be, inasmuch as they were not only—at this point, at least—discussing the interpretation of data, but whether or not to use a whole data set at all.

Collecting Nonmolecular Data
   For researchers at the GPI lab, molecular data are only valuable when analyzed in relation to other information. In Project Guajira they therefore wanted to collect and analyze sociodemographic data alongside genetic variables. They were concerned with developing the appropriate tools to account for factors such as migrations, traditions, language, and mating patterns. For this purpose, they first conducted a literature review on the region and the population, which included anthropological and historical sources. After this, the whole group participated in a three-week fieldwork trip aimed at getting to know the people and the place, obtaining the necessary approval from local authorities to conduct the research, and planning the data collection process. They chose three different research sites based on the size and composition of their populations: Riohacha, the capital of the department; Maicao, a city

Figure 5.1. Dr. Usaquén presents the project to the Wayúu (courtesy of William Usaquén, 2010).

with a large number of people called *turcos*; and Uribia, a rural municipality considered to be mainly Amerindian and "isolated."⁹ In the last site, Usaquén presented the project to the Wayúu with the aid of an interpreter and using a poster specifically designed for this purpose (figure 5.1). Through images, the poster represented the objectives of the project and the sample collecting procedure.

The poster had a map of the department with the location of the different indigenous families represented by their names written in Wayúunaiki (the Wayúu language), pictures of Wayúu people and of the researchers collecting samples, and a sign that read, "In our ancestors' footsteps." The poster constituted an effort to approach the population sensitively. The fact that it contained no technical details and that the project was explained in Wayúunaiki by a Wayúu interpreter enabled a dialogue between the GPI and the Wayúu. Nevertheless, the assumptions underlying the poster are similar to those held by other geneticists. For example, the way of representing populations created a difference between "us" and "them," even though it appealed to a common history, a common past of all of "us Colombians," which tried to erase these distinctions.

The visit was also used to pilot a survey developed to collect sociodemo-graphic and genealogical data (figure 5.2). It is worth mentioning that the GPI was the only one of the laboratories we studied that developed and used such an instrument.[10] The main purpose of the survey was to identify lineages and ancestries according to matrilineal or patrilineal descent, and to link such genealogies to mitochondrial DNA (mtDNA) and Y chromosome DNA (YChDNA), on the understanding that genealogical lineages translate into ge-netic ones. They wanted to reconstruct each individual's family tree and then use it to determine the origin of his or her possible ancestors. Interestingly, this approach to ancestries was later used in the process of creating categories by which genetic data were analyzed. The survey included information on the place of birth and family names of the studied individual and his or her par-ents, grandparents, and great-grandparents (to be filled in using a schematic family tree), language spoken, caste (the Wayúu are socially organized into castes or clans), ethnic group (used mainly for identifying the Wayúu as such), children (number, age of the mother at birth), place of residence, and date and duration of current residence.

Juggling with Categories

The construction and negotiation of categories began during the survey design stage. Before fieldwork started, the researchers developed a category called "POP" (for population) that was supposed to be filled in by the inter-viewer about each respondent (not shown in figure 5.2). It included the com-monly used population categories (indigenous, Afro, and mestizo) and two additional ones: "Arab," intended to account for individuals of Sirio-Lebanese descent, and "migrant," developed during previous research to describe indi-viduals whose parents were born in the country but outside the studied area. These categories organized people by descent but were also largely based on phenotype.[11] By using these categories, the researchers aligned themselves with widely held assumptions about the Colombian population as being tri-ethnic and about the links between physical appearance, genealogical ances-try, and genetic ancestry. Yet the new category of migrant hinted at a possible different conceptualization (although its difference from indigenous, Afro, and mestizo was not clear at the time). Thus, the survey turned out to be much more than a simple data collection tool. In it, the struggle, negotiations, and contradictions faced by these researchers seemed to materialize.

When Usaquén showed us the completed surveys, we realised that the POP category had been seldom and irregularly filled in. In some cases, "mestizo" was used when the individual was born in a specific city (outside the studied

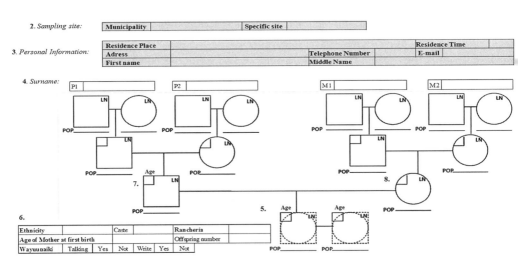

Figure 5.2. Survey developed by the GPI (courtesy of GPI 2010).

area) and in others it was used regardless of place of birth. One interviewer used the category "mestizo/indigenous" followed by a note saying, "The paternal grandmother seems to be indigenous." Similarly, the survey included a specific box for information on "phenotypic observations" (not shown in figure 5.2). The recording of these data was variable in terms of content and even whether the box was filled in or not; some interviewers used this space to write place of birth or "ethnic group," though most left it empty. Junior researcher Alonso explained, "POP was filled in according to the last name; if that wasn't clear, we used the interviewer's criteria or self-classification. [It turned out that] the mestizo was very ambiguous. It was the nonindigenous, the non-Afro. But over there [in the study area] that's not so evident."[12] Alonso explained that the categories that seemed self-evident at first, and which corresponded to the categories commonly used by geneticists, became blurred once they were filling in the surveys about actual people in everyday life. Information considered relevant and unequivocal at the beginning, in the comfort of the lab, turned out to be vague and even redundant in the dynamics of the field, where actual people were encountered. Once in Guajira, Usaquén said, they realized that

> In this region, it is impossible to find people who fit into clear-cut categories. For instance, [the category of] Afro-Colombian here is very difficult. You get some who are very light and others who are rather dark. I mean, you get the whole range of color . . . and the same goes for indigenous people, so we realized how impractical it was to think in terms of phenotype and started thinking in terms of genealogy, of ancestry . . . but then we were faced with the problem of who was a native. Here you have two types of native, those who are indigenous, but also those who have lived here forever but don't consider themselves as indigenous. So we had to divide the category into two. The same with migrants. . . . Thus, when we started thinking about this, we also realized that the categories ought to be in tune with what we were looking for. That's how we decided that depending on the [genetic] marker that we were going to study, we needed specific categories. So we came up with the indigenous-maternal and indigenous-paternal.

According to Usaquén, the difficulties of using the standard categories in practice (and the reflection that this situation prompted) had the effect of destabilizing these categories (and even the notion of category itself), questioning the assumptions behind them. To realize that people do not necessarily fit into discrete categories was a sort of eye-opener for all group members. Usaquén continued, "Yes, it was a revelation, because here you've got a mix. The process

Table 5.1. New Categories Used by GPI Researchers

Category Name	Description
Native 1	Wayúu: Parents and grandparents on maternal and paternal sides are Wayúu descendants
Native 2	Guajiro: Parents and grandparents have inhabited the region since colonial times, but do not identify themselves as Wayúu
Native 3	Wayúu-Guajiro: One parent is native 1 and the other is native 2
Migrant 1	Parents were born in the same city or department outside the Guajira
Migrant 2	Parents of different origins from outside the Guajira
Multiple ancestry	One parent is native 1 or 2 and the other is migrant 1 or 2
Wayúu-maternal	Having Wayúu mtDNA as inferred from the places of birth of mother and grandmother and their last name
Wayúu-paternal	Having Wayúu YChDNA as inferred from the places of birth of father and grandfather and their last name

of seeking and finding better-suited categories for our questions helped us understand that categories are artificial; they're purposeful and so we cannot be deterministic because we would be missing a whole lot of population possibilities. A person may belong to three different groups and if you don't have tools like the survey and are not careful with your categories, you'll totally miss that. So our label system is multipurpose and ancestral." The encounter with the field and its complexities inspired these researchers to unpack the long-standing static conceptualizations of the Colombian population. The search for a dynamic, nondeterministic system of categorizing people implied a negotiation process based mainly on the information collected through the survey. The group transformed four phenotype-based categories into eight ancestry-based ones (see table 5.1).[13]

These new categories are interesting in several ways. First, the seemingly discrete, phenotype-based categories of indigenous, mestizo, and Afro opened up into multiple and dynamic categories, which try to account for the diversity and mobility of the population in genealogical terms. For example, the different types of natives allow the specification of a group of ancestral inhabitants (i.e., *guajiros*) who are not considered indigenous and who otherwise would remain invisible. Similarly, the categories of migrants account for internal migrations that are usually overlooked. Second, by moving away from a triethnic understanding of the population, these categories avoid relying on long-held assumptions about racial markers and genetic makeup (i.e., that racialized human groups also constitute separate groups in genetic terms). Third, this system makes it possible for one individual to fit into more than one category (e.g., Wayúu-maternal and native 2), which questions the idea that population categories are simply given, static, and "natural." Last, despite the innovative character of these categories, it is noteworthy that they imply a link between genealogical ancestry and genetic ancestry, an assumption that the GPI does not question (according to this, social paternity and maternity are equivalent to the biological, and last names are considered proxies for genetic makeup).

Interpreting Data

In Project Guajira, the survey was not only a tool to collect information or to develop a new categorical system; it inspired a different viewpoint on the Guajira population that influenced the interpretations that followed. Back in the lab, in subsequent statistical analyses with different types of software, individuals were divided in advance according to the categories obtained from the survey (native 1, native 2, native 3, migrant 1, migrant 2, and multiple ancestry).[14] The researchers found statistical differences and similarities between these groups. For example, the group of individuals who had both matrilineal and patrilineal Amerindian descent (native 1, Wayúu), stood out as different from the rest; of the others, migrants (1 and 2) and native 2 became separated in a different cluster from individuals who had one Amerindian ancestor (native 3, Wayúu-Guajiro).[15] These results made the researchers regroup the six survey-based categories into two based on statistics: (a) individuals with matrilineal and patrilineal Amerindian descent (previously native 1, Wayúu); and (b) individuals with one lineage of ancestors inhabiting the region since colonial times and the other lineage of Amerindian ancestors (previously native 3, Wayúu-Guajiro), individuals who had ancestors from elsewhere in Colombia (previously multiple ancestry, migrants), and individuals who were previously native 2 (Guajiro). The group native 3 (Wayúu-Guajiro) appeared as an intermediate group in all the analyses.

These final categories used to describe Guajira's population composition were only possible because of the classificatory system previously developed and applied by the group. This means that if the group had used other categories from the beginning, the Guajira population would have been described differently; probably as less heterogeneous. What is more, the subpopulation groups were not—at any stage of the analysis—interpreted as indicating a triethnic composition of the country, and the groups were not understood as equivalent to the classic ones (mestizo, Afro, and indigenous).

The novelty of this analysis and interpretation of data is twofold. On the one hand, the GPI was able to make visible human groups that would otherwise have been fitted into standard categories. On the other hand, the researchers were capable of going beyond the self-explanatory category of mestizo. One could say that using the final two categories (individuals with matrilineal and patrilineal Amerindian descent and individuals with one lineage of ancestors inhabiting the region since colonial times and the other lineage of Amerindian ancestors, individuals who had ancestors from elsewhere in Colombia, and individuals who were previously native 2) is very similar to saying that the population is made up of indigenous and mestizos. The difference (and a significant one) is that the categories used by the GPI are not discrete and self-explanatory, as indigenous and mestizo might be. Hence, with the analysis made by the GPI, the classic categories used in Colombian human genetics research were destabilized, opening up the possibility for new versions of presenting and understanding the population.

Moving Boundaries of Innovation

In spite of the GPI's innovative, creative, and flexible process of developing categories in Project Guajira, when it came to their first attempt to enter the local and global arena of scientific production, these categories were flattened and ultimately erased. After long and sometimes heated debates on how to write what would be the first published article of the group, they decided on two things: one was to publish in English, as they considered this to be the best way to reach wider scientific audiences; the other was to divide the article in two, with one section on allele frequencies and another that articulated demographic information with genetic data (but which focused more on the demographics). This second decision was based on two motives. The first was that writing an article combining these two lines of analysis was taking more time and effort than expected and Usaquén considered that they needed to publish right away. The second was that by the time they were ready to publish, journals had stopped receiving articles on allele frequencies based on STRs; such data could only be submitted as a letter to the editor.[16] Thus, the decision to publish in English meant that they had to comply with the journals' agendas and with what editors considered relevant.

Having taken this decision and given the impossibility of writing an original article due to journal requirements, they wrote a letter to the editor reporting on allele frequencies of Guajira populations. This part of the writing process constituted a first step in stripping the categories of their innovative character, as the categories developed by the group were not used. At this point, they considered it crucial to comply with classic categories in order to increase the chances of being published. The letter was also written first in Spanish and then translated into English by someone outside the group, and the categories used by the translator were never double-checked or problematized by the authors. Thus in the process of both writing and translating, the new categories that were discussed, refined, and negotiated among all the researchers, using different sorts of technologies, suffered a major shift and became standardized. In the submitted letter to the editor, the population categories used to describe the inhabitants of Guajira were the standard ones used when publishing genetics articles on Colombian data: "Amerindian represented by the biggest *native* group. . . . *Arab* presence . . . from . . . Syria, Lebanon, Palestine and Jordan . . . and *Mixed race* group (mix of *native* with *white*)" (emphasis added). When we received a copy of this first version of the article, and after a series of conversations about it, we asked the authors for their reasons for using such categories. Junior researcher Sarmiento explained simply, "That's how these categories are in English." At that time, there was virtually no reflection on the consequences of the categories chosen in the publication of the findings, despite the negotiations that had occurred to find categories that properly suited the research objectives and findings, and the pride that they felt for the innovativeness of their work. We were struck by the fact that the classifications in the journal letter did not account for that process and, most of all, that they contradicted what the group stood for with regard to scientific knowledge and production. The categories did not do justice to the group's singular focus on reflexivity.

We would like to highlight some of the consequences of this change. It is noteworthy that the reflection on who should be considered a native is erased in the category "Amerindian." Given the way in which the category is defined, the group of nonindigenous ancestral inhabitants (i.e., native 2) is discarded. The same goes for the two categories of migrants. Not only are these categories dropped altogether and replaced by the mixed-race group, but all different migrants to the zone are blended into one group.

More interesting is the use of the category "mixed race group (mix of native with white)" as it contains both "race" and "white," two concepts to which these researchers were openly opposed. Some of their comments on the use of "white":

When you talk in terms of color, you end up adding a whole lot more characteristics and values. Like for instance that whites are the intelligent ones, blacks are lazy, and indigenous people are dumb. . . . So it's better not to think in those terms, not to use phenotype as a proxy. (Junior researcher Rocha)

When geneticists talk about Caucasian, when some talk about Caucasian mestizo, or when they try and calculate the percentages of the different components of the three ancestries a person has, all they are trying to do is to find, no matter how far back [in lineage terms] or how little [in terms of percentage], that they have a bit of white. . . . There is an obsession with whiteness, and that's so because people believe that white is better than indigenous and African. (Usaquén)

Similarly, the discomfort that race as a concept produced among the researchers required a long conversation to pin down. From the moment we met the research group, they were, like many other Colombian geneticists, against referring to human populations in racial terms. It was a visceral rejection that related time and again to the idea of the existence of pure versus impure races among humans. Our questions regarding race in genetics research were thus always received with great surprise. However, after six months of close contact, we were able to have a conversation that tackled their uneasiness with the concept. We were discussing Project Guajira, and we wanted to know what the category native 1 meant. The conversation unfolded as follows:

ALONSO: Well, native 1 means that they're pure. I mean, that the mother and father and the grandparents are Wayúu. Pure I mean for the [genetic] markers we were looking at. But we aren't looking at admixture index . . .
MFOS AND ADC: So are there pure groups?
ALONSO: No, not pure in those terms. I mean, that both parents and the four grandparents are Wayúu. . . . Not in terms of race purity that doesn't exist.
USAQUÉN: To think in terms of racial purity is a big mistake, especially because all humans are mixed. No one is pure. Yes, there are differences between humans, but to think in terms of races is wrong.
CASAS: Yes, but it is wrong because it has a history. But, as you [Usaquén] said, there are differences. I mean, humans aren't different from all other species and all are divided into races. . . . So I think that we chose not to talk about race because it's politically incorrect, and I agree with that: given the history behind "race purity" we should avoid talking in terms of races.
ROCHA: And I also think that when you do a study on present-day inhabitants of a given place you have to report all that you find, and I think that the level of

mixture is so high that clear differences can't be found, so there aren't racial differences as such. . . .

ALONSO: It depends on what you're reporting. If you want to report phenotype then you might use the classic races [white, indigenous, black, Asian], but when one reports genotype it's not possible to use racial differences.

USAQUÉN: Yes, but right now we are being politically correct and we also have to agree that there are differences between humans. . . . The problem is that when you say that and then think in terms of the classic races you're immediately attaching values to individuals of each group, and that's when things get complicated. . . .

CASAS: I would say that that's the reason why we don't use the term "race": we don't want to contribute to racism.

USAQUÉN: But it's not only that. I would say that to talk about race among humans is misleading because there isn't a real level of differentiation between humans. I mean, there is variation, but it's minor variation that doesn't mean that a given human group can be set apart in a different taxonomic group.

In the letter to the editor, the above conversation is denied and hidden, and the categories used are trivialized. In the following section we address the possible meanings of this detour and of others that took place.

A few months after the letter to the editor was submitted, we met again with the group, and during this visit we presented an earlier version of this chapter, with the aim of discussing our analysis with them. During the meeting, Usaquén expressed how, after rereading the letter to the editor, the group had realized that they were uncomfortable with the way in which it was written and with the categories used by the translator. This last point was especially annoying to them because of the fact that they had not problematized or double-checked the translated categories. In this regard, researcher Alonso said, "We were in such a rush to publish our first paper that we never even read carefully the categories; we assumed that was the way these categories were in English. That is why we have made the decision that from now on, we will write the paper in English, not let anyone else write what they want, but also that we have to explain why we use the categories that we use." The researchers said that they were considering withdrawing the article and were discussing how to include the categories that they had developed, as Rojas said, "so that it shows the complexity of our process and of our analysis."

A couple of weeks later, we visited the group again. They wanted us to see the latest version of the article, which this time had been written as a chapter of Usaquén's doctoral dissertation, to be submitted to a journal after his graduation. One of the most noteworthy aspects of this new version was that they presented an argument built upon demographic, genealogical, and genetic data. All three sources of information were given the same weight, from research design through analysis to the publication of results. They presented the survey and the role it played in data collection and analysis; they referred to the use of genealogy and demography; and they provided what can be considered a robust genetic analysis of the population structure. This new version of the article was, in many ways, a testimony to the GPI's stance and political commitment within the world of genetics. Usaquén commented:

> [Using those categories in the letter to the editor] was an unforgivable mistake. We were blinded by the idea and the rush for publishing and so we went against our own convictions. But now we know that our main product, right now, is not only publications but a method, a different way of collecting, analyzing data, and producing results. The papers will follow and we are now working on them. . . . My dissertation includes all of our papers and one more. We are ready to show our categories, our methods, and our way of doing genetics. . . . we are not afraid anymore.

This is a stance they decided not to relinquish, despite the risk of not fitting in with hegemonic standards. It is a stance that is also visible in the fact that, even though the work forms part of Usaquén´s PhD dissertation, the first author is junior researcher Rojas, as she was the one who wrote most of it.[17] Furthermore, in the final article they used the categories developed by the group: Wayúu, Guajiro, Wayúu-Guajiro, migrant 1, migrant 2, and multiple ancestry, as shown in figure 5.3.

In the article, they argue for the use of these categories and highlight the need for more comprehensive ones in genetics research, given that "traditional [genetic] classifications that coined terms such as mestizo, European, and Afrodescendant are imprecise, for the analysis of human populations requires more robust and carefully designed categories that allow more refined and adequate inferences, which correspond to criteria that show the population dynamics and historical processes of the groups under study" (Rojas et al. 2012).[18] We consider this argument to be innovative, not only because it explicitly rejects the population categories that have been stabilized, neutralized, widely used by geneticists, and mobilized in discourses of national identity, but because it also questions long-standing and widely used methods to classify Latin American populations. By doing so, they open the possibility for different paradigms and understandings of Colombian (and Latin American) inhabitants.

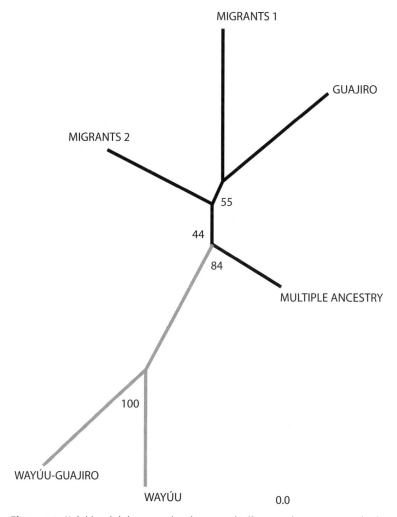

Figure 5.3. Neighbor-joining tree showing genetic distances between genealogical groups (Rojas et al. 2012).

Making Sense of the Shifts

Although there is a new version of the article, one that includes the innovative categories of the GPI group, it is worth analyzing why these researchers' approach changed so drastically between the discussions, negotiations, and subsequent writing of a BA thesis, and the drafting of the letter to the editor. In many of the conversations regarding this publication, there were two recurrent topics: the need to comply with scientific discourse and the need to follow publishing guidelines (i.e., organization of the article, kind of information provided, length, and—very importantly—language). The resulting article, indeed, looked like any other genetics publication.

One could interpret this move in terms of Latour's (1990) immutable mobiles. The writing of scientific articles is not simply a matter of expressing or translating convictions and principles. Rather, "the main problem to solve is that of *mobilization*. . . . You have to invent objects which have the properties of being *mobile* but also *immutable*" (Latour 1990: 6, emphasis in original). Thus, for findings to be mobile, they cannot be altogether innovative and new, because they will not be comparable. We could say, then, that what happened in the letter to the editor was a move toward immutability that guaranteed mobility for a group of geneticists who are, for the most part, starting their careers and need to become visible in the field. That would, however, be a short-sighted argument, as it overlooks the effort of providing new and perhaps more inclusive population categories, especially so when the group has written and will publish a different version that does not comply with the standard categories. Yes, somehow findings need to be flattened in order to travel and in order to be comprehensible to others. However, if they had not rewritten the article, these researchers not only would have presented themselves to an international community of geneticists somewhat differently from how they present themselves locally, but they would also have reproduced ideas about the Colombian population that they do not necessarily share.

In this context, our view of this group as marginal (in the terms presented in the introduction of this chapter) is helpful to understand the move from the lab to the varied versions of their article and their final decision to include the innovative categories and methods developed by them. The GPI's work goes largely unrepresented and they have openly placed themselves outside mainstream genetics. However, they still consider themselves to be geneticists and to belong to the genetics community, and they strive for a place in mainstream academia. They are what Star calls "monsters, cyborgs, impure" (1991: 29). They are doing invisible work (Star 1991) and to a great extent boundary work (Gieryn 1983), both fundamental to the discipline.

The first version of the article can be seen as an attempt at navigating within the genetics world, whereby they chose to be in tune with hegemonic stances. However, after a period of reflection, they considered that failure to include their own categories, methods, and ways of doing population genetics in their publications was more costly than the chance of being challenged by the genetics community. Regardless of the consideration of innovation as a costly endeavor, the GPI group chose to remain close to what they saw as their contribution to the wider field of genetics.

In such a process, a parallel discourse on genetics and a questioning of the taken-for-granted categories emerges. It becomes visible that "there is nothing necessary or inevitable about any such [category], all constructions are historically contingent, no matter how stabilised" (Star 1991: 38). Hence, the letter to the editor, which may at first have seemed—and in many ways was—a move contrary to the GPI's stance, but which was considered by the group as a chance to make their first entrance into the genetics network, turned into an opportunity for reflecting on their work and on the position they want to occupy within that community. In the move that took them from innovation and heterogeneity to standardization and flattening, we are able to see the impositions placed on scientific research groups. Nevertheless, the latest version of the article, and the fact that it constitutes part of Usaquén's doctoral dissertation, shows potential paths for resistance that may (or may not) turn into innovations within science.

We consider this group's reflexive work, coupled with their awareness of their role in knowledge production, to be an expression of what Anzaldúa (1987) calls *conciencia mestiza*: a consciousness that is multiple in nature, hybrid, in constant transit, and at times contradictory, but not necessarily chaotic. By its nature, it is a consciousness that needs to be self-reflexive in order to account for itself. But it is also a consciousness that "has discovered that [it] can't hold concepts or ideas in rigid boundaries" (Anzaldúa 1987: 79), and thus it needs to be flexible, to allow other ideas and postures in order to last. It requires "developing a tolerance for contradictions, a tolerance for ambiguity" (79). We found traces of *conciencia mestiza* in this group, not only in their reflexivity when thinking in terms of the kind of geneticists they are, but also in the work that they do at the lab, and even in the choices and compromises they made in publishing their work. Their hybridity and tolerance for contradiction, coupled with their marginality, might help to explain why, after a long process of negotiation, they were able to make significant compromises, only to then go back to their original and nonstandard categories and methods. The traces of *conciencia mestiza* enabled them to be simultaneously reflexive about their work and also to accept some of the trade-offs required to enter the scientific community. This enactment of genetics enabled Usaquén and collaborators to move within disciplines, and to attempt to incorporate diverse viewpoints and (at times) competing discourses, in order to do genetics as they consider it should be done, a practice that includes the need for peer acknowledgment.

Conclusion

In this chapter we have explored the practices of a particular human genetics laboratory, which provides an alternative view on genetics research in Colombia. Their reflexivity, their traces of *conciencia mestiza*, and the fact that they are marginal in many senses provide these researchers with the uncertain (though desirable) possibility of challenging and stretching population categories that are pervasive in human genetics in Colombia, while at the same time making an effort to navigate the sometimes difficult path to the academic world. The same combination of factors also makes it possible to understand how a young, small research group is capable of making contributions to the field of genetics. Their marginality and hybridity made them worth studying.

Through the example of the diverse practices and processes involved in producing a scientific article, we have elucidated how scientific politics plays a role in shaping the categories used in scientific publications (Latour 1990). In this case, the influence of scientific politics concerned the travels of human population categories from moments of flexibility and innovation toward the spectrum of stillness and inertia, and then back to innovation again. The example also enables us to bring to the surface what we call paths for resistance, which are neither final nor definitive, and which are themselves contested and mobile. This is a resistance that may eventually bring innovation and heterogeneity to scientific practice or equally may align with old ideas and stances.

Notes

1   The departments are the thirty-two political and administrative units into which Colombia is divided.

2   It is beyond the scope of this chapter to analyze these conceptualizations of people and geography, which has been done elsewhere (Olarte-Sierra and Díaz del Castillo H. 2013).

3   Haraway (1988) makes a similar point.

4   For some clear examples of contradictions that rise in genetics research see Bliss (2009a), Duster (2003b), Fullwiley (2007a, 2007b), M'charek (2005a, 2005b), Olarte-Sierra and Díaz del Castillo H. (2013), Reardon (2005).

5   This may also be due to the fact that in Colombia, postgraduate education in the field of genetics is just beginning to be consolidated. In most of the other laboratories we studied, researchers were trained abroad.

6   Other labs we studied use forensic or medical data to make population claims. Furthermore, the GPI does not use samples or data gathered or sequenced by other laboratories.

7   All the quotations in this chapter from Usaquén and his colleagues come from interviews and participant observation.

8   Haraway (1997) argues that the "modest witness" is key to early scientific discourses. These modest witnesses are untraceable white men and are modest in the sense that they are believed to be "objective." Thus the science they produce or validate is devoid of passion, interests, and tensions. Modest witnesses enable an understanding of science as the neutral translation of nature. To become unmodest is to recognize the partial objectivity of science and the undeniable role of the scientist and his or her personal history in producing knowledge.

9   Between 1880 and 1930, immigrants from Syria, Lebanon, and Palestine arrived in Colombia. Regardless of the fact that they were not originally from Turkey, people started calling them turcos, and the name became very common. Some authors say that the term reflected an initial hostility toward immigration, but later lost its offensive character and ended up being just a widely used misnomer (see Fawcet de Posada and Posada Carbó 1992).

10 While some researchers in other labs recognized the importance of nongenetic information (e.g., history, linguistics) for understanding genetic results, their tools were not as sophisticated as those used by the GPI group, and the data gathered were not necessarily taken into consideration when analyzing genetic data.

11 In this case, "phenotype" refers to physical and observable traits that have traditionally been used as racial and ethnic signifiers.

12 Comments made by researchers about the data collection process were made in retrospect during our own fieldwork.

13 The final categories, according to which data were organized and on the basis of which a BA thesis was produced, did not include a category for the so-called Turks or Arabs. According to the researchers, no one with such an ancestry (self-ascribed) volunteered a sample.

14 The software included STRUCTURE, GDA, Genepop, Arlequin, and MVSP.

15 Note that "native" has a specific definition. See previous section.

16 STR stands for short tandem repeat, repeating sequences of 2–6 base pairs of DNA.

17 It was first agreed among group members that Sarmiento should be the first author, as the project was derived from her BA thesis. She was the first author of the letter to the editor, regardless of the fact that she was a recent graduate with no additional experience. She left the group in the months following the submission, however, and Rojas, who had the same background and experience, took over as first author of the revised version.

18 This call to take into account the specificities of studied populations has also been made by other Colombian geneticists (Bedoya et al. 2006).

chapter 6

Laboratory Life of the Mexican Mestizo

Vivette García Deister

D espite its colonial origin and formerly derogatory definition, the term "mestizo" has become a diachronic convention for describing peoples in many Latin American countries (Katzew 2005). In Mexico, *mestizaje* has been assembled as the amalgam of race and nationality that confers on Mexicans their distinctive biopolitical identity (Foucault 2008). The idea of Mexico as a mestizo nation, of the existence of a "unified citizenry" (fabulated, without doubt, from a divided and unequal society), served an integrationist purpose in the postrevolutionary epoch; but this idea continues to enjoy preeminence in Mexico's regime and has made its way into a global knowledge economy. The category "Mexican mestizo population" has grown slowly but steadily since its first appearance in a scientific publication in 1954, and within the biosciences, the genome of the Mexican mestizo is deemed valuable given its differential ancestral contributions, that is to say, given its mestizaje.[1]

Since 2005, the Mexican mestizo has been the protagonist of a large-scale genomics investigation that *Nature Biotechnology* described as a "bold, race-based genome project" (Mothelet and Herrera 2005), but whose official name, the Mexican Genome Diversity Project (MGDP), stresses the search for genomic variants within the deep, horizontal comradeship of the imagined community (Anderson 1983).

This chapter deals with the inner workings of this project. Based on a laboratory ethnography carried out from May 2010 to October 2011 in Mexico's National Institute for Genomic Medicine (INMEGEN), I trace the voyage of the Mexican mestizo from the public plaza, or rather, from the university auditorium where he or she was recruited, through the wet lab and the dry lab to the database and thence to the digital information "cloud." I focus on the mediations and transformations that occur in this process and show how the relationship between *indígena* (indigenous person) and mestizo is construed at every stage. Finally, I argue that the Mexican mestizo that is currently in circulation is bioinformatically constituted, that is, he or she has suffered a material reconfiguration from blood (the sample) to bytes (information in a repository or in a cloud). This is to say that neither the blood, nor the DNA extracted from it, nor the data sets obtained by genetic analysis are proxies for the Mexican mestizo,[2] but rather that the mestizo is present in each and every configuration in a multiplicity of ways (Mol 2003), that blood and DNA and data sets and bytes are "references" that bring back the mestizo (Latour 1999).

This is also the way in which INMEGEN scientists speak of their research objects: the "buffy coat" in the freezer is not just the fraction of an anticoagulated blood sample that, after density gradient centrifugation, contains the white blood cells from which "stock DNA" is extracted; the substance in the tube is, in the actors' categories, a mestizo that they genotype and sequence in quotidian laboratory practice. The dots on a PCA graph are mestizos that exhibit specific behaviors: they cluster together or disperse; depending on their variance, they distribute closer to one reference population or to another, and in this way their behavior provides a basis for qualifying their mestizo affiliation (some mestizos may be more or less European or Native American than others).[3] But the laboratory life of the Mexican mestizo also extends beyond laboratory walls and in society (Latour and Woolgar 1986). Once inhabiting the information cloud, the mestizo takes part in negotiations of belonging that may blur national boundaries and encourage broader ethnic affiliations, therefore debilitating the protective screen of "genomic patrimony."

Blood: Sampling *Jornadas* or Crusades

On 27 June 2005, the auditorium in the School of Medicine of the Autonomous University of Yucatán, in southeastern Mexico, seated some five hundred persons—most of them students in the biomedical sciences—who attentively listened to political speeches and scientific conferences that addressed the advantages of developing medical genomics in Mexico and the reasons why they were invited to participate in a cutting-edge research project. Listeners had been attracted to the premises on that early morning by INMEGEN's well-orchestrated call for mestizo blood donors. After negotiation between Gerardo Jiménez Sánchez (INMEGEN's director general), the governor of Yucatán, the dean of the university, and the state's secretary of health, a local team was assembled to advertise the MGDP and target university students as potential donors.[4]

In addition to these efforts, Jiménez Sánchez and his research team had previously appeared on local TV shows and radio programs where they communicated in simple terms the overarching objective of their project: to understand the genetic basis of diseases that affect the Mexican population and to generate tools for managing them. The strategy worked. The first sampling *jornada* (work day) attracted not only students but also other citizens with an interest in learning about and participating in the development of a "public health genomics"—a personalized and preventive medicine tailored to the needs and features of the Mexican population (Jiménez-Sánchez 2009).[5] Outside the auditorium, colorful posters decorated the space where members of INMEGEN's research team further explained how samples would be used and processed as well as the criteria for donation. Members of the Ethical, Legal and Social Issues Unit (named after the Human Genome Project's homonymous division) collected each donor's signature and the signature of two witnesses on a consent form, and nurses from the local health system extracted the blood samples behind white screens—the only spaces of confidentiality in an otherwise media-exposed setting. Finally, members of INMEGEN's education and outreach department handed donors a commemorative T-shirt or a mug in symbolic exchange for their sample; in some cases donors were also offered a breakfast box. As a member of the education and outreach department said, "politics, science and marketing coexisted during sampling jornadas." The same modus operandi was repeated for the states of Zacatecas, Sonora, Veracruz, Guerrero, Guanajuato, and Tamaulipas throughout 2005 and 2006. During a second phase of the project, between March and June 2007, analogous sampling jornadas were also carried out in Durango, Oaxaca, and Campeche.[6]

The time invested in collecting samples (with their corresponding consent forms) and making sure they arrived safely—and refrigerated—to the laboratory where they would later be processed was equivalent to a long workday, which is why the term "jornada" was chosen to describe this chapter of the MGDP. The word also refers to a journey, which is indicative of the trip made by INMEGEN staff from Mexico City to each of the states in order to collect samples, and also of the voyage of the samples (two hundred from each state) that accompanied INMEGEN staff back to Mexico City on the same day they were collected.

The official English translation of *jornadas estatales de recolección de muestras* was "sampling crusades," which is evocative of one of the main justifications behind the MGDP. Under one of its nonreligious definitions, a crusade is "any remedial enterprise pursued with zeal and enthusiasm."[7] And sampling crusades were indeed carried out with gusto, even pride, by those responsible for them. But in what sense were they remedial? By 2005 mestizo populations had not been sampled or included in global efforts to describe human genomic diversity (such as the HapMap project in its Phase I), so it was not clear that these populations would benefit from the results obtained from such studies or have access to the products derived from them.

INMEGEN took up the challenge to eliminate this underrepresentation of mestizo populations in genomics research and to make sure that the future health benefits of genomic medicine became accessible to the Mexican people. In other words, sampling jornadas or crusades were activities anchored in what Epstein (2007) has called the "inclusion-and-difference paradigm": the inclusion of members of a group generally considered to have been previously underrepresented as subjects of biomedical research. With its focus on embodied difference, this biopolitical paradigm, Epstein argues, tends to equate group identities with medically distinct bodily subtypes in the hope of eliminating health disparities. In this way, MGDP promoters equated the mestizo identity of Mexican citizens with a measurable body: the future "genomic patient" (López Beltrán 2011).

The objective of mestizo genomics is to measure differences (across this and other groups) with regard to drug and disease susceptibility. In their first publication, MGDP researchers assessed "the benefit of a Mexican haplotype map to improve identification of genes related to common diseases in the Mexican population" and reported that their results provided evidence of "genetic differences between Mexican subpopulations that should be considered in the design and analysis of association studies of complex diseases" (Silva-Zolezzi et al. 2009: 8611). This study, they emphasized, was at the time of publication "one of the first genomewide genotyping efforts of a recently admixed population in Latin America" (Silva-Zolezzi et al. 2009). Advanced by a Mexican institute of health, this claim underscores a nation-specific response to the institutionalization of the inclusion-and-difference paradigm. Rather than emphasizing the need to represent national heterogeneity in scientific research—which occurs in other multicultural settings—INMEGEN mobilized an ideology of homogeneity characteristic of centralized nationalist discourse (Mallon 1995) and recruited blood donors under the presumption that "health research is an appropriate and important site of state intervention" (Epstein 2007: 17).

Moreover, donor recruitment for the MGDP was predicated on subject participation as a national "health good" (Petryna 2009). By ensuring that blood donation was read as a "gift relationship," where donors selflessly volunteered their time and their bodies, and making certain this effort was acknowledged on site (remember the commemorative T-shirt and mug), INMEGEN recognized the importance of establishing and maintaining public trust in the professionals undertaking custodianship of the sample collections (Corrigan and Tutton 2004). But confidence was a two-way street. INMEGEN staff also trusted the reliability of the information provided by participants. To qualify as mestizo, potential donors were required not to be recent immigrants to the country, to be eighteen years or older, to have been born in the state of recruitment, and to have two parents and four grandparents also born in the state of recruitment. The veracity of donors' declarations was neither questioned nor confirmed.

Even though INMEGEN staff explained that their project would not have immediate medical applications, project participants experienced blood donation as an event directly related to health. Before extending their arms for blood withdrawal, some donors cautioned nurses that they "had already had breakfast," worried that food consumption would somehow alter the results of the analyses their blood samples would be subjected to, as often happens with clinical blood testing. This illustrates the success of the state-mediated discourse that offered facts and statistics about how genomic medicine would improve Mexicans' health in the future—or, as Epstein (2007) has put it, the success of "recruitmentology"—in enrolling donors categorized within the largest demographic group of Mexico.

But what about subpopulations that are harder to recruit? What about those minorities said to contribute the Amerindian component of the mestizo population? Toward the end of the first phase of the MGDP, researchers sampled indigenous populations throughout the national territory (more on the reasons for doing so below). Without losing sight of the importance of establishing and maintaining public trust in research professionals, the strategy that had been successfully used to recruit mestizos was modified in order to gain access to indigenous communities. More important than political relations between INMEGEN's director general and the local governor were relations between INMEGEN staff and local sanitary authorities (rural doctors), anthropologists, and community leaders. "In fact, in these areas which were difficult to get to (one sampling site could be reached only by traveling for six hours on a dirt road or by helicopter), state and federal authorities had no leverage," commented one jornada participant. Once rural doctors had been contacted and the objectives of the MGDP had been explained to them—that is, once these key actors had been enrolled—INMEGEN staff approached the site.

We arrived in trucks that we borrowed from the secretariat of health, driven by people who knew the roads, the ranches, and the people. People recognized our drivers as those who delivered vaccines. They saw the trucks arrive and many people thought that it was another vaccination campaign. Some even fetched the kids. They understood it was a matter of health. The rural doctors have a good reputation in the communities; many times they help out not only at the clinic, but in road accidents, at people's homes. The feeling was "these people are here to help." The doctors themselves took the blood samples. Reception was good and positive. (CR participant)

This passage illustrates how different the politics of exchange were depending on whether INMEGEN staff were dealing with mestizo donors or with indigenous donors. In the first case, when sampling jornadas were projected toward members of a mestizo nation, the mediator was the state and the sampling process took place on state property. Outside the domain of the state, however, INMEGEN not only drew on the social capital of local doctors but also took advantage of existing exchanges (i.e., delivery of vaccines) to maximize the possibility of genetic inclusion of indigenous communities in the sampling process and, by symbolic extension, in the Mexican nation-state.

Set against the backdrop of a life-size poster depicting a woman with braided hair and traditional attire, indigenous donors gathered inside a small multipurpose community center in San Miguel Aloapan, Oaxaca. According to INMEGEN staff, they experienced blood donation as a matter directly associated with health by relating it, not unlike mestizo donors, to a practice they were familiar with: in this case, vaccination.[8] This occurred even as INMEGEN researchers were careful to explain the terms of indigenous participation in the MGDP, often interpreted by community leaders who acted as translators (consent forms were also printed in the Zapotec language). "They were told that they would participate in a study about similarities and differences between indigenous groups and between indigenous groups and mestizos, and that this would bring new medicine for the future, medicine that their grandchildren would be able to benefit from, and that they would be contributing to the well-being of their communities and that of the rest of the country" (CR participant). This language was successful in establishing a social relationship between indigenous donors and researchers (mediated, as we learned, by local health authorities and providers) by emphasizing their common purpose in finding ways of improving human health (Tutton 2004).

Criteria for inclusion of indigenous donors were, in addition to being born in the state of recruitment and having two parents and four grandparents also born in the state (criteria they shared with mestizo donors), that the donor and four grandparents spoke the Zapotec language. In other words, the selection of indigenous donors was based on a sociocultural criterion that is also used in population censuses to designate ethnic membership. The strategy of using language to define the ancestral component that provides genetic "singularity" to the Mexican population is consistent with a form of mestizo nationalism that "promotes an understanding of Indianness as lodged in the metaphorical gut, heart, tongue, soul, and blood of the nation and national selves" (Tarica 2008: 2).

The INMEGEN staff reported the eagerness of both mestizo and indigenous donors to participate in sampling jornadas, and donors' trust in the potential health benefits of the MGDP. Once the first results of the MGDP had been published in the *Proceedings of the National Academy of Sciences* (PNAS, in May 2009), jornada teams returned to some of the sampling sites in order to deliver them.

> We visited several states—not all of them, because we didn't have enough time. Sonora, Yucatán, Campeche, Veracruz, Zacatecas. We returned using the same contacts or with new governors if these had changed. The objective was to close the cycle, to let them know that we had acted according to the plan. We came up with a package [a colorful box that contained the PNAS paper, a DVD, and other memorabilia] and we gave presentations. One researcher talked about the way samples were processed, another one about microarray analysis. A bioinformatician spoke about data storage and we ended with the ethical, legal, and social implications of the project. We did not deliver a [genomic] map of each state, but during the presentations scientists emphasized the results obtained from the state we were visiting. It was a political-informative event. Those states that we were unable to return to received the information and the package. Donors did not necessarily attend these events. Sometimes they did: the very first donor, a man from Yucatán, did show up at the delivery of results. Some states gave this event a lot of importance. In Campeche we had university students, students from the nursing school, Maya students, military and naval authorities, the dean of the state university, the governor. Two student engineers from [Instituto Tecnológico Superior de] Calkini who had donated samples even took part in an academic exchange and spent a month at INMEGEN working with the data, their data of the indigenous population in Calkini. (CR participant)

This effort to return to the sampling sites and deliver results speaks of INMEGEN's awareness of the need to cultivate a politics of reciprocity, especially when the results of a research project are abstract and very difficult to translate—despite proficient marketing strategies—into concrete medical applications; when medically oriented genomics research is, as Eric Lander described it in his keynote address at a major INMEGEN conference, "an investment in the future."

DNA: The Mestizo Labeled and Inferred

The Genotyping and Expression Analysis Unit at INMEGEN is located on the sixth floor of a high-rise commercial office building in the southern part of Mexico City. Until 2007 when jornadas came to an end, MGDP samples were stored and processed on the fifth floor, at the so-called HapMap lab directed by Jiménez Sánchez. In January 2007 the Genotyping and Expression Analysis Unit became functional. Irma Silva-Zolezzi was designated head of the unit, and all MGDP work was transferred to the sixth floor. Automated genotyping of mestizos, according to the Affymetrix 100K SNP array protocol, was implemented. As a consequence of this reorganization, mestizos and indígenas migrated together with scientists and technicians to a higher floor of the building. This is the place where the Mexican mestizo in its DNA configuration is stored, manipulated, genotyped, and sequenced. I now turn to the labels and inscriptions used to designate—in one case—and corroborate—in another—that the mestizo has been successfully transmuted into the contents of a cryotube and is taking part in the series of transformations that objects undergo in being worked over by scientists—the matter-to-form mediations that Latour (1999) calls "circulating reference."

Sixteen digits of bar code, 3110506270300302, identify a cryotube that is kept, together with sixty-three more, in a plastic box inside a refrigerator in the Genotyping and Expression Analysis Unit. By looking at this bar code, researchers and technicians can infer that the sample in the cryotube was taken in the state of Yucatán (31), that it corresponds to a masculine donor (1), that it was collected on 27 June 2005 (050627), that the tube contains DNA (03), that it is the third sample collected in the state of Yucatán (003), and that the tube contains DNA at a "working concentration" of 50 nanograms per microliter (02). Moreover, the label on the bottom of the plastic box indicates that the enclosed samples are "male mestizos." But what is the role that these labels play in circulating reference and how do they bring the mestizo into population genomics research?

Recall sampling jornadas. After each donor's name and signature was collected on an informed consent form, donors filled out a data sheet where they provided information regarding the state of their birth, and that of their parents and grandparents. A bar code (which was lacking the last two digits of the one described above) was then pasted on the information sheet, and matching bar codes were pasted on the four tubes of blood taken from each donor. By avoiding the labeling of informed consents (which were stored separately from data sheets), there was no straightforward way of connecting the donor's identity with the bar code.

The bar code, on the one hand, anonymized the blood sample and, on the other, gave it a name that could be easily identified and circulated, one whose level of detail could be further expanded in the lab, where DNA extraction added two more digits. As one junior researcher who was involved in the design of the sampling strategy explained, the code—especially the incorporation of these last two digits—was devised to be flexible to accommodate different types of samples, not just DNA. The original aim was to create a biobank that included DNA, plasma, RNA, even cellular lines, so informed consent too was formulated with this objective in mind.[9] It was broad enough to allow different procedures and the use of samples in yet unspecified research projects derived from the MGDP (Chadwick and Berg 2001).

The label on the box is another story. Many times handwritten directly on the plastic surface in black or blue permanent marker, the label is shorthand for the type of population from which the DNA is derived: mestizo or indígena, male or female. For those states where only mestizos were sampled, the label can be inferred from the first three digits of the bar code. But for those states where both mestizos and indígenas were sampled, the label on the box contains additional information and thus may read, "Mayas 01 Yucatán." The Maya qualifier is not codified in the bar code but enters genomics research all the same. Researchers expect mestizos to present genetic evidence of mestizaje, while they expect the contrary from Maya samples. Let us see how these expectations are dealt with in laboratory practice, and how the social categories mestizo and indígena appear in genomic science.

Sample 3110506270300302 and many others are subjected to PCR-based or DNA sequencing genotyping. The objective is to assign a mitochondrial DNA haplogroup and a Y chromosome DNA haplogroup to each sample. The most common mitochondrial haplogroups in America are A, B, C, and D (Torroni et al. 1994; Maca-Meyer et al. 2001); one of the most common Y chromosome haplogroups in continental Europe is R, which is widely distributed across the Iberian peninsula (Karafet et al. 2008). If the sample presents haplogroups B and R, for example, "this indicates mestizaje."[10] Practiced on dozens of samples every day, in all the cases that I observed, genotyping invariably confirmed that the label "mestizo" is correct, and in this sense genotyping provided only qualitative evidence that a sample classified using the social category mestizo is in fact—genetically—a mestizo.[11] On the possibility of genetically defining mestizaje, Silva-Zolezzi commented:

> You can do it at the level of mitochondrial and Y haplogroups in males. If one individual has indigenous mitochondrial and European Y, you can define him as mestizo. Now, you may come across an individual in which both haplogroups are European or both haplogroups are indigenous, so at this level you cannot define him as mestizo. . . . I have several of these cases. . . . Imagine an individual: you obtain the sample, and you don't know his history. You don't know his name. You don't know who he is, and you run his mitochondrial haplogroup and his Y chromosome haplogroup. As it turns out, in Oaxaca you find that 60 percent of your male mestizo population presents indigenous mitochondrial and indigenous Y, just like any indigenous group . . . and there you could not say whether this individual, if you've only measured his mitochondrial and Y haplogroups, you cannot say whether he is [genetically] mestizo or indígena.

The solution to this conundrum, in which a sample may be classified socially as mestizo and, upon first examination in the laboratory, genetically as indígena, is to analyze nuclear DNA.

> [We address the problem] by using nuclear DNA via markers that we know present differentiation between European and indigenous groups—and a substantial number of them, because if I take only five nuclear markers for an individual who presents indigenous Y chromosome and mitochondrial haplogroups, the chances that all five of them will turn out to be indigenous are very high—if I test for one hundred markers I am going to find out that this individual is not indígena: he is mestizo because he starts to present European markers. And if I run a thousand markers and then fifteen hundred, I will fine-tune these [ancestral] contributions.

The search for European markers underscores the resilience of the social category mestizo, which, once entering genomic research (in the sampling process), is not only difficult to drop but actually fuels practices of corroboration. When applied to indigenous samples, this fine-tuning of ancestral contributions also opens the possibility that in one individual designated as indígena by the sampling strategy and corroborated as such by genotyping, a European ancestral contribution may be detected via ancestry informative markers (AIMs).

If treated as evidence of admixture in the same way as in the previous case, the label on the sample and the individual's group affiliation would change, based on statistical inference, from *indígena* to mestizo. Silva-Zolezzi explained why this situation is possible: "I know of no indigenous population that is 100 percent Native American or, in quotes, 'pure' [*simulating quotes with her fingers*]." She continued to explain that indigenous populations may present more or less degree of admixture. From all the indigenous populations sampled by INMEGEN, she added, "Mayas have the highest degree of admixture. . . . What happens is that in indigenous populations the detectable amount of admixture is smaller than the one detectable in mestizo populations." The reason to distinguish between indígena and mestizo and, more importantly, the reason to retain the label indígena in the face of genetic admixture is described thus: "You continue to include them in the indigenous group because that is the way they were defined . . . because indigenous populations are not only genetically indigenous; they are indigenous in culture, in usages and customs. They self-describe themselves as such and this correlates with their genome . . . but there are individuals within these groups that carry genes and variants from a European genome . . . generally in a smaller proportion than the average mestizo." Those individuals who cluster closer to mestizos than to their indigenous sample population are called outliers by the scientists because they are atypical; they are numerically distant from the rest of the data set and fall outside the average distribution of their samples. Because they do not confirm the indígena/mestizo divide, scientists tend to exclude them as "genomic noise." Given the relatively high degree of admixture of Mayas from the state of Yucatán, outliers appear with more frequency in this group. But there are also mestizos that practically overlap with indígenas; this has happened with samples from Oaxaca, Campeche, and Guerrero. Even when these samples statistically behave like indígenas, their behavior is interpreted by MGDP researchers as pointing to "different qualities of mestizaje."[12]

The effect of these interpretations—that, upon closer examination, the indígena may be considered to be admixed while the mestizo inevitably remains present (albeit in different material configurations)—is the reiteration of Mexico as a mestizo nation. Mediated by nuclear DNA analysis, the indígena is once again genetically assimilated into the mestizo. The mestizo lives experimentally latent in the indígena, who resists the force of mestizaje; but there is no turning back from mestizo to indígena.

Data: Mestizo Metrics

After being genotyped by the hundreds, the mestizos travel from the Genotyping and Expression Analysis Unit to another facility on the sixth floor: the Supercomputer and Information Technology Unit. They arrive at a Linux cluster once again recodified, this time into binary code. But the life of the mestizo in this cold, fully air-conditioned room inhabited only by two computer analysts and a stuffed penguin (the Linux mascot) is secondary to the attention it receives back on the fifth floor, at the Computational Genomics Lab. This "dry" lab is a medium-sized office with four work stations equipped with PCs and eighteen- to twenty-inch monitors linked to the cluster.

Of the four researchers working there, only Juan Carlos Fernández is fully dedicated to the bioinformatics of the MGDP. Fernández's work involves data curation, analysis, and visualization—the marks of data-intensive science (Hey, Tansley, and Tolle 2009). He spends many hours of the day "getting to know" MGDP data, extracting patterns from large data sets (for which his training in artificial intelligence has proved useful) and performing dozens of queries before running ancestry analysis software or producing a graph he can discuss with Silva-Zolezzi.[13]

The materials and methods section of INMEGEN's publication in PNAS can be roughly divided into two parts: sample collection and wet lab procedures (discussed above), and dry lab procedures. The latter involve the use of EIGENSOFT software, which includes the algorithm EIGENSTRAT (Patterson, Price and Reich 2006). This software is based on a statistical approach called principal component analysis (PCA). As explained by Fujimura and Rajagopalan (2011), EIGENSTRAT is used to divide samples into groups or clusters based on the similarity of their SNP (single nucleotide polymorphism) variation scores. Although EIGENSOFT was created to correct for population stratification in genome-wide association studies (Price et al. 2007), it is particularly useful for building clusters via algorithms "that do not depend on pre-labeling or pre-sorting samples . . . nor do they require pre-specifying the number of expected clusters or groups" (Fujimura and Rajagopalan 2011: 11).

By modeling SNP variation scores (which MGDP scientists regard as ancestry differences) along continuous axes of variation, the program allows scientists to visualize dots on a graph: data sets that represent clusters (which MGDP scientists regard as populations) from which inferences can be made, based on their distribution (e.g., based on their ancestry, mestizos are very far away from Africans and closer to Europeans than to East Asians).[14] How were these informatic tools used to visualize and account for the relationship between Mexican mestizos and indígenas?

Despite the tenacity of the indígena/mestizo divide as a way of classifying the Mexican population for genetic studies (see chapters 3 and 7, this volume), sampling of indigenous donors in addition to mestizos was not initially put into practice. The need to collect indigenous samples was made explicit by PNAS referees in the second half of 2008. Referees asked that the study should include a contrasting population that would help to make sense of the designation used for the populations studied.

The indigenous qualifier, which translates into the category Native American in scientific publications, is a common currency used in the United States to distinguish the descendants of pre-Columbian inhabitants from people of mixed descent. For the team at INMEGEN, during a first attempt at publication (where they reported results obtained using EIGENSTRAT technology) all that was not East Asian, African, or European and that, according to the graphs resulting from their principal component analysis, corresponded to a fourth—initially unlabeled—ancestral population, was inferred to be Native American. Reviewers for PNAS required confirmation that the behavior of the data obtained from mestizo populations (that they distributed along an imaginary line between Europeans and another "pulling" vector) was in fact attributable to a population that INMEGEN scientists could label as Native American.

One of the reasons why indigenous populations were not sampled from the outset of the MGDP was to avoid potential conflicts with these groups. Throughout the twentieth century, indigenous populations in various regions of Mexico were sampled for scientific purposes—mostly by medical geneticists aided by physical anthropologists (Suárez and Barahona 2011). The experience of the Human Genome Diversity Project (HGDP), which in the process of collecting blood from indigenous groups across the world faced resistance, conflict, and scandal (Reardon 2001), was well known at the time the MGDP was being planned. One way to avoid a similar fate was to move away from "anthropological questions" like the ones raised by the HGDP, and to promote instead a "strong biomedical focus" such as the one championed by the more auspicious HapMap project (Dennis 2003).[15] Another way was simply to avoid sampling indigenous groups. By modeling the MGDP on the HapMap project, INMEGEN employed the first strategy, but could only postpone the more politically sensitive decision to collect and process samples from indigenous populations.

In 2005 Jiménez Sánchez met with state governors and with local and federal health authorities to evaluate the possibility of sampling some of these sites. In good measure because Oaxaca's governor Ulises Ruiz was keen on supporting INMEGEN's incursion into this territory, the state of Oaxaca and its numerous indigenous population became a viable possibility.[16] When PNAS referees required inclusion of a Native American group, Zapotecs immediately came to mind. As Silva-Zolezzi told ethnographer Ernesto Schwartz two months before the article was published, sampling Zapotecs was "a pragmatic decision" that made use of a state-mediated concession.[17] In this way, INMEGEN's sampling strategy for the MGDP was modified through a series of negotiations with both methodological and political overtones.

Thirty unrelated Zapotec individuals from a community in Oaxaca's Sierra Norte were designated as the Amerindian ancestral population in the analysis of genomic diversity in Mexican mestizos. Remaining in a space of difference not granted to other Amerindian population samples (for, as I showed in the previous section, all other indigenous populations were ultimately considered to be admixed), Zapotecs became a reference population (ZAP) in a way similar to CEU, YRI, JPT, and CHB populations of the HapMap.

Regarded as a pragmatic scientific decision, this move was justified in the article by citing the results of an article published in the *American Journal of Human Genetics* in June 2007. This article indicated that "the Zapotec from Oaxaca in southern Mexico provide the best predictor for the Native American ancestry of LA Latinos (19%) and Mexicans (18%)," and it favored "the Zapotec as the single most useful population for the purpose of building a map [of admixture]" (Price et al. 2007: 1029).[18] But ascribing Zapotecs a different space and time is in agreement with twentieth-century mestizo nationalism, which, supported by state *indigenismo*, was successful in assimilating indigenous cultures by regarding them as primitive or at least unsynchronized with respect to the rest of the population (Tarica 2008; see also chapter 3, this volume).[19]

Unaware of this ideological likeness, researchers produced a graph that shows mestizos as distributed between—but clearly separated from—Europeans (CEU) and Native Americans (ZAP). Native Americans were first inferred (they were the "other" default ancestral population), and then enrolled—in a somewhat subsidiary way—as the population that gives Mexican mestizos their distinctive statistical behavior. Indígenas are thus "placed at the pre-historical origin of the nation, at its metaphorical roots" (Tarica 2008: 2). Simultaneously, results of the genomic analysis suggest that genetic differences among Mexican mestizos are mainly due to differences in their Amerindian contributions, therefore promoting "an awareness of shared time and space between Indians and non-Indians, the core of the mestizo nation" (Tarica 2008: 2). Mestizo genomics can, accordingly, be seen as a practice of both differentiation from and identification with the indígena.

Over the past few years there has been a growing demand for samples of "Mexican ancestry" to be included in ambitious whole-genome sequencing projects, which aim to provide comprehensive resources on human genetic variation and tools for further computational analysis. The 1000 Genomes Project includes seventy such samples, while Complete Genomics has sequenced five of them. All of these samples are subsets of the population sample of people with Mexican ancestry created for the renowned HapMap3 project. This project includes samples of people with Mexican ancestry from Los Angeles in addition to ten other populations. It provides data about population-specific genetic differences that are meant to be useful for genetic association studies. Because the risk of some diseases of major incidence among the U.S. population has been correlated with certain patterns of admixture (e.g., a higher risk of diabetes in individuals with a mix of Native American and European—i.e., "Mexican"—ancestry[20]) and the U.S. Census Bureau has identified Hispanics/Latinos as the fastest-growing ethnic group in that country, the need to include mestizo samples in genomic analyses has been widely recognized.

For Fernández, there is an additional motivation for including them: "Latinos and mestizos are in vogue because availability of samples translates into availability of data." Moreover, he adds, "innovation is not in the methods of computational genomics, it is in the samples." That is, whatever matter-to-form mediations the mestizo undergoes in computational analysis, only good samples make good data sets. This sample-to-data transformation is where the value of INMEGEN's mestizos resides, especially for foreign interests.

Mexico's recently reformed general law of health indicates that, in order to protect "genomic sovereignty" (Benjamin 2009; Schwartz-Marín and Silva-Zolezzi 2010), human samples collected in Mexican territory must be associated with an ongoing research project approved by a Mexican institution (namely, INMEGEN). Moreover, designated as a national reference center and a consultative organ of the federal government, permission for samples to leave Mexican territory is subject to approval by INMEGEN and exportation must abide by a laborious legal protocol. This process can discourage sampling and potential inclusion of Mexican mestizos in international genome projects.[21] But what better way to gain access to the highly valued (and not so readily accessible) samples of mestizos from Mexico than by obtaining access to MGDP data? In the rest of this section I describe a collaborative project in which MGDP data—from both indigenous and mestizo population samples—played a key role. As I illustrate below, access to information about mestizos supervened on access to information about indigenous samples, which raises a number of issues.

Admixture mapping is a method commonly used for localizing disease genes in admixed populations, where it is necessary to distinguish between potential associations due to ancestry and those due to disease haplotypes (Mao et al. 2007; Tandon, Patterson, and Reich 2011). While a large number of AIMs is required to discriminate between both kinds of associations (or to "correct for stratification"[22]) with a degree of precision, commercial SNP arrays can be used to extract sets of AIMs that allow researchers to infer ancestry in mestizo populations.

A multi-institutional project to design a panel of 446 AIMs useful for admixture mapping and haplotype-disease associations in Latin American populations has been completed (Galanter et al. 2012). The object of the project was to create "data infrastructure" that will be useful not only for Mexicans but also "for Latinos in the entire American continent" (Fernández, personal communication, 19 October 2010). The development of a database that contains a large enough but select subset of AIMs useful for "correcting association signals" will allow each researcher to "choose the AIMs of her preference on the basis of her experimental design and investigation" (Fernández, personal communication, 8 October 2010), be it for correlation with "normal phenotypes," association with disease, or correction for population stratification. In other words, the AIMs panel will decontextualize facts about Mexican mestizo populations so as to facilitate their travel, and at the same time enable their recontextualization for use in new research settings (Leonelli 2010).

The panel includes markers that estimate three major ancestral components of contemporary Latin American populations (African, European, and Native American) and can distinguish populations of different regions, which according to the authors renders the selected panel of AIMs "broadly portable to populations from throughout the Americas" (Galanter et al. 2012). INMEGEN contributed genotype data of samples derived from three indigenous groups: twenty-five Mayas (southeast), twenty-one Zapotecs (center), and twenty-two Tepehuanos (north). Data from other indigenous population array samples (Aymara from Bolivia, Quechua from Peru, and Nahua from Mexico) derived from other collaborators' samples and data sets. In addition to these six Native American ancestral populations, four ancestral populations from the HapMap3 project were also used (CEU, TSI, YRI, LWK), plus another one from Spain.

One feature of the material transformation of samples into genotype data is that they tend to travel in groups (Leonelli 2010). The "most informative" Native American data are grouped together with data from other ancestral populations (not just those from Mexico). Then, by using the AIMS panel to genotype samples of Mexican mestizo populations and comparing the resulting ancestry estimates with those provided by genome-wide association studies (as developed in the MGDP), the evidential value of the panel is confirmed. This process of "validation," as the authors call it, is crucial for rendering the AIMS panel useful for genotyping not only Mexican mestizos, but also mestizos throughout the Americas. In this material configuration, mestizos are once again bolstered by indigenous qualities, but these are no longer enclosed within the limits of the nation. As Fernández put it, "Sometimes we think of Mexico within the frontiers, but this [AIMS panel] should work for all of Latin America" (personal communication, 23 September 2011).[23]

Two conclusions can be drawn from this example. First, the existence of the mestizo in the information cloud facilitated an international and multi-institutional collaboration that resulted in a publication by forty-six authors from twenty-eight different institutions and ten countries (Fernández is listed as second author). This draws attention to an "ethos of collaboration" in the making (Leonelli and Ankeny 2012). Collaborators tend to share data instead of samples. In the case of INMEGEN, custodianship of MGDP samples and data is separated. Despite this apparent separation, the quality of data (i.e., which data sets Fernández considers to be appropriate for use) is evaluated on the basis of the conditions in which samples were obtained, as well as on the way in which samples behave statistically. The existence of signed consent forms and information about the collection site (e.g., geographical location) is important to establish validity about a sample's provenance, but just as important is statistical placement of any given sample within a sample population (what Fernández calls "context"). That is, knowledge about data inevitably entails knowledge about samples, thus preserving epistemic continuity between them.

The second conclusion is that participation of MGDP data in research on admixed populations throughout the Americas has consequences for the biosociality (Rabinow 1996) of the Mexican mestizo. The AIMS panel resulting from this collaborative effort is envisioned as one that will provide "useful resources to explore population history of admixture in Latin America and to correct for the potential effects of population stratification in admixed samples in the region" (Galanter et al. 2012: 4). The journal *PLoS Genetics* was selected for publication on account of its "philosophy of open access and wide dissemination," which resonates with the authors' goal "for the AIMS panel to become an open standard that can be used by researchers throughout the Americas in future genetic studies."[24]

Previously confined to the political margins of the nation, the mestizo now reaches out beyond these borders and boasts new representativeness: he or she is no longer Mexican, but also Latino/a or Hispanic. Equating nationality and mestizoness was crucial to recruitment of blood donors and to the establishment of a nationalist research agenda. It is only when the mestizo/a reaches the information cloud that her or his ethnic affiliation is broadened (and, in a way, the category is destabilized) so that he or she actively plays a part in transnational research agendas (see chapter 7, this volume).

Conclusion

Influenced by twentieth-century nationalist discourse, the relationship between mestizo and *indígena* inside INMEGEN's laboratory is construed as one in which the indigenous body contributes genetic qualities that make the mestizo both a unique representative of the Mexican nation and a unique object of biomedical investigation. By focusing on mestizo transformations, I have shown this to be true at every stage of the research process, from the sampling of blood to the sharing of data. The peculiarities of the mestizo/indígena relationship are used as grounds for regarding mestizo blood samples as "genomic patrimony," a national good that must be zealously guarded from foreign attention. But those same peculiarities make mestizo data useful for biomedical investigation across Latin America, therefore loosening the grip of "genomic sovereignty." From INMEGEN's collaboration with other American research institutions (especially those in the United States) there follows a concerted effort to ascribe the mestizo a supranational racial and ethnic identity. The series of mediated transformations described throughout this chapter makes room for these expanded affiliations. This new episode in the laboratory life of the Mexican mestizo contrasts with the nationalist discourse of INMEGEN in the early days, when the objective was, rather, to disperse (from the center of Mexico) an integrated biomedical account of the Mexican that put the interests of the nation before the registers of region, culture, or ethnic group (Mallon 1995). The laboratory life of the Mexican mestizo/a points in the direction of his or her identitarian revision under a scientific lens of undeniable social meaning.

Notes

1   The purpose of the 1954 study (published in the *American Journal of Physical Anthropology*) was to test the applicability of a photoelectric reflection meter to assess small differences in average skin color in Mexican communities with varying degrees of mixture between Amerindian and European ancestors (Lasker 1954). A subsequent article, published in 1963 by Rubén Lisker, a Mexican medical geneticist, reported hereditary hematological characteristics in both indigenous and mestizo groups. From this point on, the study of hemoglobins and antigens was common until the 1990s, when the possibility of analyzing the genetic structure of mestizo populations shifted researchers' focus from these molecules to larger sets of genetic markers.

2   Montoya (2011: 90) argues differently for the research practiced on the "Mexican diabetic" along the Mexico-U.S. border. What he calls "the diabetes enterprise" begins with the premise that "the individual biologically represents the community, that DNA represents the individual, and thus that DNA is a proxy for the population of donors of [the area he has named] Sun County."

3   Principal component analysis (PCA) is used to produce a graphical representation of clustering and dispersion in a set of variable data.

4   At that time the governor of Yucatán was José Patricio Patrón Laviada; the dean of the university was Dr. Raúl Godoy Montañez; and the state's secretary of health was Dr. Jorge Sosa Muñoz.

5   The term "public health genomics" is the result of a conference held 14–20 April 2005 in the Rockefeller Foundation's Bellagio Study and Conference Center in Italy. A group of eighteen participants gathered there with the purpose of establishing an international network to promote the goals of public health in the light of new knowledge about the human genome and the development of a novel range of technologies. An outcome of the Bellagio Initiative was a model for public health genomics that aimed at securing "the responsible and effective translation of genome-based knowledge and technologies for the benefit of population health" (Bellagio Initiative 2005: 7). Although not explicitly adopted by INMEGEN directives, this model provided a general framework for the institute's overall goals and activities.

6   A methodological note is in order. On 25 October 2010, a group of five people collectively narrated a story that helped me to bring sampling jornadas into being. Given the extension of jornadas (four years, 2004–2008) and the life of the MGDP (eight years, 2004–2012), the collective narrative that emerged was composed of participants' partial work and life histories, of their testimonies as privileged witnesses of jornadas, and of memories. Because INMEGEN had photographically documented jornadas, we used archived photographs as "memory sites" for recollection (Gray 2002). When structuring the focus group, which I dubbed "Collective Remembrance" (CR), I deliberately allowed participants to propose the course of action (prompted by images) in order to reconstruct past events, and interrogated them during the process. Participants in CR complemented each

other's recollections; sometimes they corrected each other and supported these corrections with detailed descriptions and visual evidence—this is how remembering was socialized and collectivized. Most of the ethnographic data used in this section were collected during this process.

7 Webster's *New International Dictionary* (unabridged), vol. 1, 546.

8 According to the journalist Silvia Ribeiro (2005), some of these sampling expeditions were linked to programs of social assistance such as Oportunidades (see http://www.oportunidades.gob.mx/Portal/), which suggests that the association between INMEGEN's project and other ongoing programs for disease prevention in rural communities was intentional.

9 Although the biobank project never materialized, INMEGEN is still a member of the Public Population Project in Genomics (P3G), which is dedicated to facilitating collaboration between researchers and biobanks working in the area of human population genomics.

10 The result is also consistent with the founding myth of the Mexican nation: that all Mexicans descend from a European (more specifically, Spanish) father and an indigenous mother.

11 Female samples (which lack Y chromosome DNA) are subjected to a different process. Given that mestizaje in these samples cannot be inferred from only their mitochondrial haplogroup, female samples undergo nuclear DNA analysis with a set of highly informative ancestry markers. The cost and degree of resolution of this analysis varies depending on the number and type of markers used.

12 Elsewhere (García Deister 2011) I have analytically described two categories of mestizaje. There is a discrete mestizo (either one is or is not mestizo) and a continuous mestizo (one can be more or less mestizo), both of which play different roles in scientific practice. Qualitatively defined, the first one usually confirms that the labeling of samples is correct, while the second one, defined in terms of percentage of ancestral contributions, allows comparison of individuals or subpopulations.

13 Even though INMEGEN was conceived as a state-of-the-art genomics research center that embraces data-intensive science, how much of the data management is taken to be the main part of the scientific enterprise and whether a bioinformatician can be listed as a major contributor, or even first author, in a publication is many times contested. Fernández's impression is that only recently has the importance of his work been fully appreciated: "Giving biological meaning to a piece of data is relatively easy; the difficult part is extracting the data—knowing what parameters to use [in order to retrieve information]—this is the role of the bioinformatician. . . . I am the first one to visualize the data. I see them even before the biomedical researcher does" (personal communication, 21 September 2010). As I argue below, the role that Fernández plays in collaborative projects derives from his ability to manage data.

14 For a critical analysis of how these kinds of plots are interpreted as representing genetic ancestry differences, or as a consequence of the "translation or slippage

from clusters of genetically similar samples to categories of samples with . . . 'shared ancestry,'" see Fujimura and Rajagopalan (2011).

15 Scientists in the MGDP often expressed a need to confer on their research a well-demarcated disciplinary identity. The main concern was to establish that their questions could be inscribed within biomedicine or applied medical genomics and not anthropology. During a conversation I held with Silva-Zolezzi on 5 May 2010, she verbalized her concern that, once she left INMEGEN to work in Nestlé and other researchers came to occupy her post (at that time it was presumed that these would be anthropologists), the project would "become more populational and less focused on the search for clinical application." On 26 June 2010, Silva-Zolezzi further explained that "anthropologists tend to look for uniqueness in human populations, which is interesting from a historical or evolutionary point of view, but this does not explain differences in terms of disease between human groups. Common variation, which is the object of MGDP, is useful for understanding disease." In this way, Silva-Zolezzi associated the search for common variability with medical genetics, and the search for rare variants with anthropological genetics.

16 A member of the Partido Revolucionario Institucional (PRI), Ulises Ruiz served as governor of Oaxaca from December 2004 to November 2010. He is well known for having dabbled in corruption since the beginning of his office (for which he was dubbed "Ulises Ruin"), and is especially noteworthy for his unlawful treatment of members of the Sindicato Nacional de Trabajadores de la Educación, who faced up to Ruiz regarding pay raises and other labor-related issues. On 14 June 2006 he ordered the eviction by force of some 70,000 protestors from a central plaza in the city of Oaxaca (which they managed to reclaim after seven hours of confrontation). The Asamblea Popular de los Pueblos de Oaxaca (APPO) was formed shortly afterward in response to Ruiz's actions. The APPO accused Ruiz of political fraud and acting with impunity, and demanded his immediate resignation. Despite the Senate's recommendation that Ruiz step down from office (followed by a similar one by the Chamber of Deputies), Ruiz refused to do so. He completed a six-year office under what the Senate described as "serious conditions of ungovernability." With the blessing of Ulises Ruiz, INMEGEN signed on 22 March 2007 an agreement with state health services and educational institutions to sample "ethnic groups" in the region. This event did not go completely unnoticed. The journalist Silvia Ribeiro (2005) depicted the MGDP as a "vampire project" (a term originally used for the HGDP) that would be useless to the indigenous people of Oaxaca but would in turn benefit researchers, institutions, and transnational pharmaceuticals.

17 In December 2008, a participant of jornadas told the ethnographer Ernesto Schwartz that despite the support granted by local authorities (or perhaps because of it), retrieving the samples from Oaxaca was not an easy task. The connection between INMEGEN and the infamous Ulises Ruiz set off alarms, and the APPO and other indigenous organizations came close to demanding that the samples

be returned to the community. According to Silva-Zolezzi's account, the first in-digenous jornada in Oaxaca proceeded smoothly and yielded good results, which prompted extending the analysis to other regions.

18 This type of decision—which population to designate as a reference for Native American ancestry—is not set in stone and may change during the course of a research project. In a conversation we held in May 2010, Silva-Zolezzi said, ret-rospectively, that she would no longer use ZAP as the reference population for Native American ancestry because they do not conform a genetically homogenous group.

19 Separating the Zapotecs in space and time is what anthropologist Johannes Fa-bian (2002) has called "the denial of coevalness."

20 See Montoya (2011) for a thorough analysis of research on diabetics of Mexican ancestry in the United States.

21 Samples have been shipped from INMEGEN to laboratories outside Mexico for very specific projects under this legal protocol, nonetheless. On the other hand, research groups in other institutions (including state hospitals) have continued with genomic research on human samples as usual. This responds in part to the fact that such research groups predated the creation of INMEGEN. Moreover, investigation in these places is more clinically oriented (e.g., diabetes) and the research site guarantees privileged access to blood samples (i.e., patients and controls attending the hospital).

22 See the introduction for an explanation of the technique of controlling for popula-tion stratification, which allows researchers to match cases and controls in terms of ancestry, thus allowing them to discriminate between variants that are associ-ated with disease and those that are a product of demographic and evolutionary processes.

23 This assertion reminds me of a remark made by Leopoldo Zea, a prominent Latin Americanist, in one of his last works: "Frontiers can no longer be extended be-cause everybody is within them. . . . This is what we are living in that long frontier of Latin America called Mexico" (2002: 13).

24 Letter of the authors to the editors of PLoS Genetics, dated 4 January 2012.

chapter 7

Social Categories and Laboratory Practices in Brazil, Colombia, and Mexico

A Comparative Overview

Peter Wade, Vivette García Deister, Michael Kent,
and María Fernanda Olarte Sierra

Genetic analysis of the ancestry of white Brazilians and of a sample of Colombians conducted respectively by Sérgio Pena and Gabriel Bedoya—and colleagues—show quite similar results: an overwhelming majority of European contribution to the Y chromosome; significant African and/or Amerindian contributions to the mtDNA; and more than 75 percent European contribution to the autosomal DNA. Yet the research teams diverge in their interpretation of these results. As different categories are emphasized and reified, their interpretations result in images of the populations under study that are significantly different. While in the Brazilian case Sérgio Pena and colleagues emphasize the admixed character of the country's population, Bedoya and colleagues emphasize European descent (Bedoya et al. 2006; Pena et al. 2009; R. V. Santos et al. 2009). Both interpretations, however, resonate well with dominant ideologies in their respective countries about the constitution of the national population. As discussed in other chapters, in Brazil such ideologies strongly value mixture. In Colombia, while dominant representations cast the country as a mixed one, there is also a strong tendency to see the nation as consisting of several distinct regions, some of which (among them the region sampled by Bedoya et al.) are explicitly associated with strong European origins. The work of Pena et al. in Brazil and Bedoya et al. in Colombia is not, of course, exhaustive of genetic research in each country. However, the contrast between the two studies makes it possible to discern alternative possibilities that may materialize in a particular area of genetic research. As we aim to show in this chapter, this offers an important avenue for interrogating the production of genetic knowledge, as well as for exploring the multiple interactions with social categories and understandings over the course of a project's trajectory.

Building on the material already presented in this volume, this chapter addresses the difficult question of how categories of human diversity operate in the practice of genomic science in the laboratory. The focus here is quite specific, exploring the research process in the laboratory, from agenda setting through interpretation and publication of results. A broader comparative analysis of the contexts in which this occurs and the issues raised by such an analysis are the subject of the conclusion.

The discussion of science studies in the introduction brought out two key themes. First, social categories do not simply enter the lab and shape practice there, undermining the reliability of the results produced by scientific methodologies; instead, the kinds of categories we are exploring—of race, ethnicity, region, population, nation—are already natural-cultural assemblages, co-produced by diverse actors and materialized in objects and practices over long periods. Our aim, therefore, is not to argue that social categories contaminate scientific results. It is rather to explore how natural-cultural categories operate in scientific practice and produce certain effects: the focus on one particular topic, rather than another; the inscription of one category, rather than another; the highlighting of one ancestral link, while placing less emphasis on others. This relates to the second theme, diversity. Scientific practice is diverse, by its nature: scientists disagree and contend over the best methods and interpretations. For them, this is inherent in an ongoing process of achieving greater and greater reliability in their results. Diversity of results also responds to particular scientific contexts and aims: for scientists it is normal that what made sense at one time and for one purpose might not work so well in another context. Social studies of science can use this diversity to explore what is at stake in particular ways of working, as opposed to others; how some methods and interpretations become standardized, while others do not; what gets taken for granted when some practices become standard.

In this chapter, we build on the diversity we encountered both between countries and within them. The spread of the project across three countries and across different labs within the same country allowed us to access interesting multiplicity in everyday scientific practice and in the interpretation of data. We use this to show how things could have been different and how this difference reveals the ways in which natural-cultural categories work in geneticists' research.

Categories and Latin American Lab Practices

Before we turn to the scientific practices themselves, it is worth reviewing briefly the kinds of categories that are at issue in the genetic practice that was our focus. The categories are natural-cultural entities that have been formed by circulation through varied expert and nonexpert discourses and practices; they generally have considerable historical depth and thus often predate the practice of genetics. In the scientific publications and in lab practices we explored, the key categories for referring to populations in English—the main language of scientific publication—are African (including related categories such as African-descent, African-derived, and African Colombian or Afro-Colombian), European, Amerindian or Native American, and mestizo or admixed populations (understood as the product of sexual reproduction between the continental categories).

National categories are frequent: Mexican, Brazilian, and Colombian, as well as others. Regional terms are also quite common, referring to overarching spaces such as Latin America, South America, and so on, and specific areas within countries (often administrative divisions, but also ones that might carry strong cultural and historical associations). Categories such as indigenous may also appear, while the terms "black" and "white" occasionally appear in print, used mainly by Brazilian and occasionally Colombian scientists.

Spanish and Portuguese terms such as *indígena, negro, preto, pardo,* and *blanco* or *branco* occasionally appear in Spanish- and Portuguese-language publications. Apart from "indígena," these terms tend to be used only in texts published by Brazilian researchers, in part because some of the terms are used in the national census—"preto" (black), "pardo" (brown), "branco" (white). However, in the labs, such terms are used more frequently—for example, in interviews with geneticists and in everyday conversations among geneticists themselves.

As described in the introduction, there are significant differences in the ways *mestiçagem* or *mestizaje* has worked in Brazil, Colombia, and Mexico. This has consequences for the everyday taxonomic categories that are used in each country. In Mexico, a dualistic taxonomy that divides indigenous and mestizo is dominant. In Brazil different classificatory systems coexist—the three main census-based color categories (branco, pardo, preto), multiple everyday taxonomies that include terms such as *moreno* or *mulato*, and a more politically inspired system that encourages an inclusive black (negro) identification, opposed to white (branco). There is a tension between a tendency to classify in oppositional terms—white versus black—and a tendency to classify with multiple and flexible categories. In contrast to this assimilation of pardos with pretos, in Colombia mestizos are more likely to be terminologically assimilated to whites. The white/mestizo categories then tend to be set against blackness (*negros, morenos, afrocolombianos, comunidades negras,* or black communities) and indigenousness (*indígenas, indios*). These differences in taxonomies, simplified here, need to be borne in mind when exploring what happens to population categories in the practice of genomic science.

Defining Agendas: Mestizos and Nations

As is common in genomic research on human diversity, one important rationale for the research projects we analyzed was phrased in terms of making a contribution to the global project of discovering the genetic causes of complex disorders.[1] Other important rationales for research were the elucidation of population and migration histories (e.g., Alves-Silva et al. 2000; Wang et al. 2007) and the creation of databases for forensic genetics. In Brazil, an additional aim for some was to "uninvent" race as a concept and critique models of human diversity based on the concept of population clustering (Pena 2007, 2008).

Within these broad agendas, among the geneticists we studied, the clearest ways in which race-like and national categories appeared in the whole lab enterprise were, first, in the initial choice to focus on admixture and its ingredient populations, and second, the frequent use of a national frame for the sampling and analysis. We deal with mixture and mestizos first and pick up the question of the nation later on.

The Mestizo Nation and Its Ancestral Components

The focus on mixture often meant an interest in mestizo populations as the quintessential products of Latin American admixture, although it necessarily entailed an interest in the populations seen as close to the three founding populations of the original admixture: African-derived, indigenous, and European-derived populations. In Colombia, Emilio Yunis was a good example of a primary focus on mestizaje (Sandoval, De la Hoz, and Yunis 1993; Yunis Turbay 2009). For the labs we studied in Medellín and Bogotá, mixture or the Colombian mestizo were an important focus of analysis (e.g., Bedoya et al. 2006; Carvajal-Carmona et al. 2003; Rey et al. 2003; Rojas et al. 2010; Torres Carvajal 2005). For some geneticists, this obeyed a medical rationale. One said, "Here you have a hybrid, a mix, and you can see how diseases behave differently depending on each person's ancestral composition" (interview). Others had a more general interest in population: "It is important to study the mestizo to know the current genetic composition of the country's population; to see what we are, because we are all mixed . . . but we have to see the behavior of that mix because it is different; it has to be different [in different places]" (interview). Geneticists in Colombia also sampled Afro-Colombian and indigenous populations (Builes et al. 2008; Rojas et al. 2010; Rondón et al. 2008; Yunis et al. 2005), but a key focus was on the process of admixture and the diversity it produced.

In Mexico, scientific interest began to shift away from Amerindians and toward mestizos in the 1940s. This partly obeyed an apparent decline in the percentage of indigenous people, from 29 percent of the national total in the 1921 census, which used racial categories (*razas indígena, mezclada, blanca*), to 14 percent in the 1930 census, which counted speakers of an indigenous language, a more restrictive category.[2] In line with this, INMEGEN's entire initial project was to explore the genomic diversity of the Mexican population, defined in advance as mestizo (Silva-Zolezzi et al. 2009). For Irma Silva-Zolezzi, the genome of Mexicans was a valuable object of study because of the differential ancestral contributions, where genetic information derived from the three founding populations of admixture has been conserved over time, across different generations.

In Brazil, this focus on mixture was particularly marked, as noted in chapter 1, and it was also directed in part at dismantling the very idea of race. "In Brazil, we are all *mestiços*" was a frequently heard statement in genetics laboratories in that country. Or, as Sérgio Pena framed it in his exposition during the public hearings of the Supreme Court on racial quotas, "practically all Brazilians have the three ancestral roots present in their genomes." Earlier studies on racial mixture tended to take the idea of race itself rather for granted (e.g., Franco, Weimer, and Salzano 1982; Ottensooser 1944, 1962), even if the emphasis on admixture was strong.[3] In the research of Pena and Bortolini, in particular, intensive mixture in Brazilian populations became proof that there is no scientific basis for the very idea of race (Pena and Bortolini 2004; Pena et al. 2000). They used the standard census color and race categories, but in order to show that they are not valid biological ones.

While geneticists had written about genetics, race, and society before (e.g., Salzano and Freire-Maia 1970), the more recent arguments were being constructed in close connection with social and political debates about the relevance of racial categories for the design of public policies. In these arguments, Brazilian mixture was given positive connotations. As Bortolini said in an interview, "Miscegenation is good, because the alternative is segregation. This is why it is important to show the results of our research to the public." Like Pena, she conceptualized genetic mixture as an antidote to racism and the use of racial categories.

Although mixture was an important element in the research of many geneticists in Brazil, revealing the admixed character of the Brazilian population was generally not as explicit nor as politically motivated as in the research of Pena and—in lesser measure—Bortolini. As Sidney Santos of the Universidade Federal do Pará expressed it, "The mestiço is like the white of your eyes, everyone has it" (interview). Showing that Brazilians are all mestiços, for Santos, was not a primary object of genetic research. He was more interested in processes of migration and in the diversity in processes of admixture that had occurred in Brazil (Leite et al. 2009; N. P. C. Santos et al. 2009). For him, as for many geneticists in Colombia and Mexico, the mestiço was the unmarked, commonsensical category.

Black and Indigenous People in the Mestizo Nation

Research that, instead of focusing on mestizos, studied white, black, and indigenous populations might paradoxically end up reinforcing the dominant image of the mestizo nation. In Brazil, for example, people who classified themselves as branco (white) were the focus of studies by Sérgio Pena and his colleagues, but the outcome was to show that they had mixed genetic ancestry (e.g., Parra et al. 2003), thus reinforcing the emphasis on mixedness. This is linked to the way mestiçagem works as a discourse of national formation (see introduction, this volume). The discourse is all about mixture, but it necessarily involves constant reiteration of the foundational categories of Amerindian/indigenous, African/black, and European/white. The original categories are simultaneously set apart from the nation (they are not mixed), and included in it (they originally helped give rise to mixed people; they can still become mixed now through assimilation into the mixed nation). The way genetic research deals with indigenous and black people reflects this, although in different ways.

Studies of Native Americans—of which there is a long and complex tradition (Salzano and Callegari-Jacques 1988)—could create a distancing effect: the emphasis was on these populations as distinct and separate from the rest of the nation, albeit in a process of transition, including admixture (Coimbra et al. 2004; Sans 2000). Indigenous people tended to appear as an object of genetic investigation distinct from mixed people. This fits into long-standing tendencies to treat indigenous populations as relatively isolated and genetically distinctive, if not pure (Barragán 2011; Santos 2002). Salzano and Bortolini (2002: 327), in their overview of Latin American population genetics, decided not to focus on Amerindian populations at all, as these had been the subject of recent overviews that focused entirely on this topic. Indigenous people were a separate subject. In one sense, this simply obeyed the objectives of the research (for example, the peopling of the Americas, or microevolutionary dynamics), but it also reinscribes the place of indigeneity as something separate and distinct within the (mixed) nation.

Some research also focused on indigenous peoples as ancestral or parental populations, located implicitly in the pre-Columbian past, rather than the present, even though the sampled populations are present day (e.g., Rojas et al. 2010; Wang et al. 2007). In the case of INMEGEN, a Zapotec sample was used as a proxy for the ancestral past.

Research on black populations often dealt with them in terms of both their difference and their mixedness. In Colombia, sampling of African-descendant, African-descent, and Afro-Colombian populations marked them as distinct from mestizos and yet also mapped their diverse ancestral admixtures (Builes et al. 2008; Paredes et al. 2003; Rojas et al. 2010). More recent studies of African-descent quilombo dwellers in Brazil separated them out as specific objects of research, while often noting levels of admixture (e.g., Palha et al. 2011; Ribeiro et al. 2009). One study noted that several genetic studies of quilombos in Amazonia have focused on "evaluating the level of isolation and mobility, and quantifying the gene contribution from non-blacks among them" (Guerreiro et al. 1999: 163). The early work of Bortolini analyzed samples from black populations in "partially isolated villages" in Brazil and Venezuela, emphasizing the mixed ancestry of these populations, which made them distinct from African populations (Bortolini et al. 1995; see also R. V. Santos et al. 1999; Schneider et al. 1987). Other black populations were sampled in big cities, such as Porto Alegre and Salvador, and results showed that those classed as pretos (blacks) were mixed, yet they were still presented as part of a separate "African-derived" category (Bortolini et al. 1999). The tendency to treat black populations as separate was less marked in Brazil, where blackness is seen as a much more important part of the national mix, but it was nevertheless evident when dealing with people seen as located at the most African end of the color/race spectrum.

These differences in the treatment of indigenous and black populations reflects long-standing differences in how indigeneity and blackness fit into discourses of mestizaje and its "structures of alterity," in which indigenous people figure as the "other" much more firmly than black people (Wade 2010: 37). In genetic research, indigenous people are seen as distinct and relatively unmixed; mixture turns them into mestizos—the category "Amerindian-derived" population almost never appears. Black populations are, by definition, African-derived or -descent. By implication, they are mixed, as the genetic data confirm—just as "everyone" is mixed—yet this does not make them fully into mestizos. A separation is made in both cases, but for indigenous people it derives from the perception of them as small groups encysted within national society, culturally and geographically separate, while for black people it derives from a perception of them as more integrated into national society, yet still defined by phenotypical difference.

In one sense, this overarching emphasis on mixture seems unexceptional: it is "only natural" that genomic research in Latin America should have this focus. It has been a long-established trend in Latin America (Salzano 1971; Sans 2000); there is a broader medical interest in admixed populations as useful objects of investigation for tracing significant genetic variants (Bliss 2008; Darvasi and Shifman 2005); Latin American populations are perceived as mixed; and there are many reasons for wanting to understand the population history of the Americas. In short, admixture studies are, in many ways, quite a stable assemblage.

Unsettling the Mestizo Nation

Yet the stability of the assemblage is not quite all it seems. All representations of mixture in Latin America are characterized by a tension between sameness and diversity (Wade 2005). Everyone is mestizo (except apparently some indigenous groups—on which more below), but this unity is differentiated by a host of distinguishing classifications, most fundamentally by the three categories held to be the founding origins of the mix—black/African, white/European, and indigenous/Native American. Recently, multiculturalist reforms have put into sharper relief the question of sameness and difference, driven largely by recognition of black and indigenous rights. The genetic studies we explored reflected exactly this tension: while they focused on mixture and/or mestizos, they also inevitably presented diversity, which challenged the stability of the mestizo category. How much mixture did it take for an indígena or an Afro-Colombian to be a mestizo?

The stability of the mestizo nation was also questioned by the depiction of regional variety, so that different regions or subunits within the nation had greater or lesser amounts of European, African, or Amerindian heritage. The natural-cultural category of mestizo, as well as being perpetually stretched by the founding categories of European, African, and Amerindian, was split and refracted by natural-cultural categories such as region.

Another revealing aspect of the tensions underlying the mestizo/mestizaje assemblage was the double emphasis given to whiteness. On the one hand, even people who considered themselves white in Brazil, or who had 70–80 percent European ancestry in Colombia, or who were labeled Caucasian mestizos, had high levels of indigenous and/or African ancestry in their mtDNA

(Carvajal-Carmona et al. 2000; Yunis et al. 2000, 2005).[4] So they were perhaps not quite as white as they thought. On the other hand, some studies of the autosomal DNA of mestizo populations, above all in Brazil and Colombia, showed a high level of European contribution; analysis of Y chromosome DNA showed high levels of European ancestry too (Carvajal-Carmona et al. 2000; Pena et al. 2009, 2011).

The dual emphasis on mixture and whiteness is not a contradiction in genetic terms because the mtDNA analysis looks back to early colonial encounters, while the autosomal DNA results show the effect of more recent events— the time frame for each result is different (R. V. Santos et al. 2009: 799). Yet the point remains that the genetic research could show different aspects of mixture. While the genetic process of mestizaje—and its result, the genetic mestizo—commanded wide acceptance as a focus of study, it nevertheless evinced the tensions characteristic of mestizaje, highlighting difference as well as similarity. National populations were mainly mixed and thus characteristically Latin American, with distinct indigenous and black margins, but also plenty of European ancestry—which could be understood by nonexperts as meaning they were quite white.

The stability of mestizaje as an assemblage is also questioned by the work of the Expedición Humana in Colombia in the 1980s and 1990s, which sought to unveil hidden aspects of Colombia, both genetic and cultural, with its focus on isolated populations and particularly indigenous and Afro-American groups, although occasionally also on Colombian populations (Ordóñez Vásquez 2000; see also chapter 2, this volume).[5] The diversity associated with black and indigenous groups was highlighted by this project in ways that arguably contrasted with the dominant emphasis on mixture and the mestizo nation. Of course, the Expedición Humana became the target of controversial and polemical accusations of colonialist "gene-hunting" by some indigenous activists, anthropologists, and geneticists (Barragán 2011).[6] It also ran the risk of portraying indigenous and black groups as marginal (distant, isolated, etc.) in relation to the unmarked but dominant mestizo nation, without necessarily paying attention to the unequal power structures that construct such marginality.

But alongside these critiques, another reading of the Expedición Humana is possible, which recognizes that the overall image of the nation it produced unsettled the dominance of the mestizo image simply by focusing on contemporary indigenous and black populations, in a way that arguably placed them at center stage and did not portray them only as exemplars of difference and isolation or of the inevitability of mixture. In some ways, the Expedición was a manifestation of previous and current ways of doing genetic research, which tended to focus on isolated populations—which laid it open to polemical accusations. Yet there is no doubt that its basic rationale included challenging the taken-for-granted mestizoness of Colombia; this is indicated by the participation of activist anthropologists such as Nina de Friedemann and Jaime Arocha, who contested the invisibility of Afro-Colombians, pressed for multiculturalist reform, and wanted to recast the whole image of the nation. The Expedición thus hinted at ways of doing genetics otherwise, differently from a dominant emphasis on mixture.

The Nation as Organizing Frame

The second major feature that defined agendas in many of the labs we studied was the use of a national frame to organize the research. In Brazil and Mexico, the genetic character of the nation was an explicit object of attention and was linked overtly to national policy issues around affirmative action and/or public health. Pena (2002), for example, coined the term *Homo Brasilis* in order to express the distinctive cultural and genetic character of Brazilians, reflecting earlier attempts to do the same—for example, Freire-Maia's (1973) *Brasil: Laboratório racial*. A national frame was also evident in the work of Emilio Yunis in Colombia, especially in his more popular writings in which he pondered the fate of the nation, asking, "Why are we like this?" (Yunis Turbay 2009). The nation was also the explicit context for forensic genetic research, which aimed to provide a tool for the identification of national citizens, thus building up a kind of biological citizenship (Paredes et al. 2003). One team in Bogotá aimed to build a national database, centralizing disparate information "to have the broader picture of the Colombian population [based on] the allele frequencies of the different populations of the country."

The genetic diversity maps that INMEGEN produced for Mexico and that forensics projects produced for Colombia "naturally" showed the nation-state, with national boundaries implicitly suggesting genetic difference. In the case of Mexico, this emerged explicitly in discourses of genomic sovereignty, which suggested a particular quality to Mexico's genome (Benjamin 2009; Schwartz-Marín and Silva-Zolezzi 2010; Schwartz 2008). In Brazil, too, discourses of genomic sovereignty were evident, building on ideas of the particular character of the national population's ancestral admixture: the high degree of African ancestry, the diversity within the country, and the noncorrespondence between social color/race categories and genetic ancestry have, for example, been seen to require specific applications of pharmacogenetics and forensics (Pimenta et al. 2006; Suarez-Kurtz 2011). In short, then, the natural-cultural category of the nation entered into the lab as an organizing device and was reproduced there.

However, the category of the nation was also destabilized; there were clearly other ways of organizing genomic science that worked with different categories. First, as already noted, some geneticists chose to focus on specific regional populations, seen as possessing interesting characteristics. The Antioqueños (often known as *paisas*) in the northwestern highlands of Colombia and the Gaúchos of southern Brazil are two prominent examples. For both regions, strong regional identities are attached to these populations, which emphasize their cultural difference and their high levels of European ancestry. One of the local geneticists leading the research in Medellín stated: "Paisas' genetic composition is almost unique in the country, because of the kind of mestizaje they had and because the paisas were isolated for many, many years. . . . Among the paisas the European racial component is very, very high; it is the highest in the country." There were also strictly technical reasons. The Antioqueños were seen as a useful population because they were said to have a specific history, characterized by initial intensive mixture followed by relative isolation (an argument we examine later): "optimal application of this approach [of admixture mapping] in Hispanics will require that the strategy used is adjusted to the specific admixture history of the population from where patients are being ascertained" (Bedoya et al. 2006: 7238). In other words, regional populations could work more effectively than national ones as objects to circulate in international genetics research.

For Brazil, the importance of the regional profile of the Gaúcho at the Federal University of Rio Grande do Sul is highlighted by the different research done at the Federal University of Minas Gerais. As seen in chapter 4, the population of Rio Grande do Sul has a very distinctive regional identity. In contrast, one of the coauthors of the *Retrato Molecular do Brasil* viewed the regional identity of Minas Gerais as closely connected to Brazilian identity: "Here in Minas [Gerais] people say that the *mineiro* is the synthesis of the Brazilian. We have even seen this in genetics: he is a third African, a third European and a third Amerindian."[7] When asked why research had not been conducted on the genetic profile of the mineiros similar to that on the Gaúchos, the researcher quipped: "It's because the Gaúcho wants to be different; the mineiro just wants to be Brazilian." Thus, the category of region could operate in contrasting ways, both reinforcing and undermining the image of the nation.

Second, the transnational character of genetic science—the networks and the publication strategies, the circulation of data derived from samples—meant that some researchers explicitly adopted a broader frame, situating themselves in Mesoamerica, South America, Latin America, or the Americas (Bortolini et al. 1995; Martínez-Cortés et al. 2010; Wang et al. 2007, 2008). International networks transcended national boundaries. Several researchers in all three countries were involved in CANDELA (Consortium for the Analysis of Diversity and Evolution in Latin America), a multinational research project coordinated by Andrés Ruiz Linares, a Colombian geneticist at the University of London. Researchers from INMEGEN had links with Stanford University and the Genome Institute of Singapore, while many Brazilian geneticists worked in international collaborations. This suggests that nation was not a category that could circulate and work as effectively in transnational science circles as admixture or mestizo, which were genomic objects that could yield interesting results. The national origin of a given population seems to have been less functional in this respect, although it still mattered to specific scientists who were interested in saying something about the character of their nation, whether in terms of medical or population genetics. For example, during the INMEGEN-Stanford collaboration, INMEGEN researchers emphasized the need to identify indigenous variation common to Mexican mestizos and potential medical applications for the national population, whereas those based at Stanford focused on identifying transnational indigenous variation from a more demographic and comparative genomics viewpoint.

In sum, it is clear that the genomic research reviewed here focused on specific themes. Mestizaje and the mestizo were in one sense an obvious choice and constituted a stable device for organizing research. However, the way these processes and categories unavoidably pulled in diversity and heterogeneity—centered around Afrodescendants, indigenous people, whiteness, and regions—made them a less stable assemblage than appears at first sight. The nation was also an obvious frame in some instances, but regional foci and transnational agendas provided other motivations. Overall, the effect of these choices is to create the strong impression of mixed populations, which are generally located in nations, but sometimes in continental regions, and which have a good deal of European ancestry, especially in certain subregions. Indigenous and African-descent populations may be made visible, but in ways that reiterate both difference/isolation (especially for indigenous people) and mixedness (especially for blacks) and may sometimes locate these populations in the past.

Sampling People: Categories, Maps, Roots

The definition of agendas already shapes sampling, in the sense that it defines which populations one is going to target. But the details of how sampling is done can shape results in more specific ways too. A number of themes emerged in this respect. The first concerned how categories were chosen and used to classify samples: these reflected attempts by geneticists to generate meaningful genetic results using nongenetic categories. The second concerned how sampling choices reproduced (but also in some cases altered) a sense of the distribution of biosocial populations in space, creating maps of genetic diversity that often bore indelible marks of the decisions made about how to sample that diversity. The third was about how populations were made to appear rooted in space (and time) by the decision to sample individuals seen as properly native to the locality by virtue of their genealogical parentage.

Categorizing Samples: Continuity and Innovation

In terms of the categories used to classify sampled populations, the standard ones used in the genetic research we studied were, as noted earlier, mestizo, African-derived, indigenous or Amerindian, and, in Brazil, the census categories of race/color. Indigenous populations would often be subclassified by ethnic or linguistic group. National categories might be used, or implied, and regional or other territorial subcategories were also often used. In Brazil, the wide variety of categories and taxonomies circulating in everyday social life was reflected in a concomitant heterogeneity in the practice of classifying samples in genetic research (R. V. Santos et al. 2009; Santos and Maio 2004).

Overall, the standard categories were familiar to geneticists, their scientific audiences, and the wider public. In that sense, they tended to reinforce familiar ideas about population diversity and how to describe it. However, on occasion, new categories emerged in the design of projects: the way these arose and changed is an interesting illustration of how categories act and are acted on in the laboratory and what happens to categories that depart—more or less—from standard portrayals of population diversity.

One example was the discussion around race/color categories to be used in the Brazilian arm of the international CANDELA project. This project sought to sample people in five countries and relate their biocontinental genetic ancestry (African, European, Amerindian), aspects of their phenotype (e.g., skin tone and morphology), their self-classification according to a defined list of terms, and their perception of their own ancestry in terms of African, European, and Amerindian components. There was a good deal of discussion around the set of race/color terms from which respondents would choose to classify themselves, partly because a single set was meant to cover different countries.[8] In Brazil, there already existed the national census terms of "preto, "pardo," and "branco." But pardo is a very broad category, and the geneticists in the Brazilian team wanted to use the category mulato. One of them explained, "Mulato has black traits; pardo is a dark person who does not have African, black traits. Pardo is lighter than mulato. . . . Here it is socially easier to say that one is white than black. So if one already says one is mulato, then . . . I think it is possible to place the mulatos and negros in the same category, for more precision." The view was that people would be likely to avoid identifying as negro in the questionnaire (preto was not available as an option), as this is a highly charged term, potentially being either an insult or a statement of racialized political affiliation. The researcher said, "In Brazil, black people have a lot of prejudice about skin color. A negro classifies himself as pardo. This is a big problem for geneticists."[9] It was seen as a problem because people with "African, black traits" (and high percentages of African ancestry) would be in the same category as people of much less African appearance and ancestry. If mulato was available as an option, then the pardo category would split between more and less African individuals and thus match better with what the researchers thought their genetic data would show. When analyzing the results, they would also be able to merge the negros and the mulatos together as the categories with the most African phenotypical and genotypical traits.

Within a complex system of social classification based on race and color, researchers were trying to reduce the way perceived social bias in classification practices might affect the results of genetic analysis, and to make categories of self-classification align more closely with genetic realities as they saw them. These race/color categories were natural-cultural assemblages, to the long history of which genomics was adding the twist that mulato could be seen as a biologically more accurate category. This represented the reinsertion of a category that was quite common in genetic racial mixture studies in Brazil in the 1960s–1980s—including the subcategories of *mulato claro* and *mulato escuro* (light mulatto and dark mulatto)—but which had been largely abandoned from the 1990s onward (Azevedo 1980; Krieger et al. 1965; Salzano and Freire-Maia 1970). In the case of CANDELA, a category that was nonstandard in terms of official Brazilian race/color classifications was brought into the research and successfully made to stick—or rather, stick again. It had a genetic logic behind it and it was already a familiar (if unofficial) category.

In the case of the Colombian project analyzed in chapter 5 by Olarte Sierra and Díaz del Castillo, a category system that was otherwise had a more checkered career. One of the labs in Bogotá designed and ran a project that sought to explore the genetic diversity of a specific region of the country, including an indigenous group. During the project, which collected biological samples and data on individuals' genealogical and geographical origins, the team came up with a series of ad hoc categories for classifying individuals in terms of these data. There were three categories of natives (*nativos*) depending on paternal and maternal origins (indigenous and local nonindigenous), two categories of indigenous people classified by maternal or paternal indigenous ancestry only, two categories of migrants (*migrantes*) from outside the region, and a further category of multiple ancestry (*ascendencia múltiple*). The categories emerged in an attempt to make sense of the sociological reality of the population, as encountered by the researchers, but, as in the Brazilian case, this was also felt to accord with some expected genetic results: as the lab director said, "When we started thinking about this, we also realized that the categories ought to be in tune with what we were looking for. That's how we decided that depending on the [genetic] marker that we were going to study, we needed specific categories."

However, when it came to submitting the results of the research to an international English-language journal, these multiple categories ended up being simplified to much more familiar categories, such as Amerindian and mixed race: a translator selected these terms and the researchers accepted them without problematizing the choices. An article using these categories was submitted and rejected—and later the team wrote a new article self-consciously opting to return to their own category system. The uncertainty over taxonomies and the inertia that allowed classic categories to reassert themselves suggests the fragility of trying to proceed otherwise and reiterates the fairly durable and stable assemblage based on the mestizo (or rather the admixed population) and the Amerindian as devices in genomic science. Unlike the category mulato in the Brazilian case, the categories the Colombian lab came up with were innovative and were displaced by translator's terms assumed to be neutral and standardized, with initially no discussion about the implications of the displacement, even though they later decided to reinstate the categories they had developed.

In both these examples, researchers classified their sample populations with nonstandard categories, with the aim of generating interesting results that might make social identities more congruent with expected genetic facts than the standard categories. In one case, this move has endured, because the category was already a very recognizable one. In the other case, the tactic did not, initially at least, make it past the first hurdle of translation into English, thus demonstrating both the fragility of categorizations of population diversity that are otherwise—that is, orthogonal to the standard descriptions—and how powerful the standard categories are.

Samples in Space

The second theme in this section concerns how sampling choices related to the distribution of biosocial populations in space and the production of maps of genetic diversity. Projects often chose standard administrative territorial divisions to organize the sampling. The research that sought to define a Colombian map of allelic frequencies for the purpose of forensic identification used the country's administrative departments (Paredes et al. 2003; see also chapter 2, this volume).[10] The INMEGEN project used the Mexican states (Silva-Zolezzi et al. 2009). Sérgio Pena's work used the five major regions into which Brazil has been officially divided since the 1940s and which are widely used in scientific and everyday representations (Pena et al. 2000). Researchers at the Federal University of Pará in particular conceptualized Brazil as genetically differentiable in its macroregions. As one of them put it, "The regions of Brazil are very different. Every region has a different history of miscegenation, so their populations are different." These geographical entities were then reproduced on maps and were characterized in terms of genetic profiles, for example showing the percentage of biocontinental ancestral contributions to each population, which seemed to imply that these were naturally distinct entities. Of course, administrative subdivisions are not simply random: they often follow geographic and historical factors that have influenced demographic processes. But they are not natural units either. They are natural-cultural hybrids par excellence, which enter the lab as obvious frames for organizing samples and are reworked there, this time with a genetic dimension, in ways that usually reinforce dominant and popular knowledge about the racialized ancestry of the regions.

In the Colombian case, the taken-for-grantedness of this sampling technique was laid bare and defamiliarized in a revealing way. By using the standard departmental map to organize their sampling, the researchers ended up with a regional division of the country that was quite at odds with the conventional one. The latter usually shows five regions—Andes, Pacific, Caribbean Coast, Orinoco, and Amazon.[11] The Pacific region is usually perceived as a geographical, historical, and cultural—and racial-ethnic—unity, being seen as the black region of Colombia. Its northern half is a single department, Chocó (in which 82 percent of the population identified as Afro-Colombian in the 2005 census). Its southern half, however, cuts across three different departments. Paredes et al. (2003) used a statistical process to produce clusters of departments, according to allelic frequencies. This produced four major regions, in which the southern half of the Pacific region was swallowed up into what they called the southwest Andean region, for the simple reason that their regions—and samples—followed departmental boundaries and thus could not show the Pacific region as a single entity. Sampling procedures reproduced the department as a natural-cultural entity, but in doing so unsettled the natural-cultural entity of the region. The regional geography of the nation ended up being otherwise—but only as the unintended artifact of sampling procedures that were being carried out as normal at a lower level of geographical resolution. This indicates the power of proceeding as normal when it comes to sampling.

Rooting Sampled Populations

The third theme related to sampling procedures concerns the common practice found in many of the genomics projects we observed of selecting individuals whose parents, and preferably four grandparents, had been born in the location of the same taxonomic sampling unit as the selected individual— whether that was a Mexican state, the Mexican nation-state, a Colombian geographic region or municipality, or the Brazilian Pampa region (Bedoya et al. 2006: 7238; Guardado-Estrada et al. 2009: 696; Marrero et al. 2007: 161; Paredes et al. 2003: 67; Silva-Zolezzi et al. 2009: Supporting Information, 1). When the sampling unit was an indigenous group—for example Zapotec in Mexico or Wayúu in Colombia—we found that parents and grandparents would be required to speak the relevant language, have the appropriate surnames, or have been born in the right place to ensure authentic indigeneity (all these data would be supplied by the individual being sampled). The purpose of this practice—common in genomics projects of this kind—is to create a sample understood to be rooted in the spatialized population that is under study.[12] The idea is to avoid including the descendants of anyone who migrated to the sampling locality in the last thirty to fifty years—even if such individuals might be deemed Zapotec or Wayúu in the eyes of local people. Limiting the criterion to grandparents is a pragmatic recognition of the limits of people's memories, but it also assumes that, if researchers go back two generations, they are able to access some kind of populational homogeneity, albeit of a provisional kind. The practice conveys an image of people traditionally settled and isolated in geographical and demographic niches, only disturbed by migrations in the last three to five decades, as if migration and movement were not likely to be a constituent aspect of social life before this. Of course, it is likely that movement and migration will have increased markedly in the last few decades. However, the practice of filtering the sample population according to geographical ancestry renders less visible the whole process of population movement; it also presents whatever locality is under study as homogeneous across its spatial extension and over decades-long periods of time. It takes the definition of the population out of the hands of local people and employs geneticists' criteria (Reardon 2008).

The rooting effect was particularly noticeable in the research on the population of Antioquia, explicitly identified as a "population isolate" (Bedoya et al. 2006). A large part of the argument about isolation focused on individuals with great-grandparental ancestry that rooted them in six small municipalities, identified as founding settlements, located in a subregion of Antioquia, known as Oriente. The aim of the research was to tap into a core, original paisa population, seen as distinctive culturally, historically, ancestrally—and genetically. Yet the paisas are not coterminous with the Antioqueños, as the latter refers to anyone born in the department's large area, which extends to the Caribbean coast in the north and the major river valley of the Cauca to the east, where many dark-skinned Antioqueños live. Paisa usually refers to the lighter-skinned, highland population, descendants of precisely the Oriente populations sampled in the genetics project. The sampling process reinforces this paisa definition of Antioquia and the whiteness of its population, which underwrites the idea of the paisas as particular and special. The researchers chose to sample in this way with the express purpose of getting at the roots of a particular population, perceived as distinctive. In the context of their project, this aim is perfectly justifiable. In doing so, however, they inevitably reproduce an image of that population as particular. A project that chose to sample Antioquia right across its geographical coverage would have come up with a different picture of the region, as might a project that did not choose to limit its sample to people with great-grandparents born in the region or locality.

Alternative practices were possible: in Brazil, for example, some studies used the sampling technique of filtering by grandparental ancestry, while others did not, depending on the objectives of the study. Research on the Gaúchos of southern Brazil—identified as a "distinct population" (Marrero et al. 2007: 160)—rooted their samples in this way, as did research by Sidney Santos at the Federal University of Pará on indigenous people and on Afrodescendant residents or quilombos. Other work undertaken by Santos revealed the specific ways sampling strategies resulted in different population profiles. Research in the city of Santarém, Pará state, northern Brazil, compared rooted and unrooted samples in order to assess changes wrought by the migrations that had strongly affected the region since the 1950s. The rooted population sample ("natives of Santarém," with all four grandparents born in the town) had higher proportions of Amerindian and lower rates of African ancestry than the sample representing the total population of the town (Santos et al. 1996: 511, 514). Although both populations showed high levels of admixture, one kind of mixture was constructed as more authentic—or native—than the other, and as more representative of the original process of mixture thought to have occurred in the Amazon region. More recent work by Santos has focused on the creation of a national genetic database for forensic purposes (Ribeiro-Rodrigues et al. 2011).[13] This has specifically avoided the historical-geographical homogenization created by rooting samples: as researchers wanted to establish a profile of the current population—including recent migration flows—rather than the original populations, the criterion of place of birth of parents was not applied in the selection of individuals for sampling. This contrasts with Colombia, where rooting practices were used in projects aiming to create national genetic maps for forensic purposes.

Thus different sampling practices make visible multiple populations as biosocial objects: where some notion of native origins—often with notable racialized dimensions—was at stake, the population appeared as rooted and genealogically authentic. Where there was an interest in movement and mobility, a different kind of population appeared. In all the sampling processes examined in this section—using descriptive categories, using geographical units, choosing individuals to be sampled—the effects of standard practice are defamiliarized and made more evident by looking at ways of proceeding otherwise, even if these ways are often fragile. At the same time, the power of normalization is also made clear.

Interpreting Data

The way samples are chosen and defined influences the kind of interpretations that will result, just as starting agendas will also shape sampling and interpretation. These stages of the research process, addressed separately here, overlap and interpenetrate. Here, however, we want to focus on how meanings are attached to the data generated by the samples.

The Antioquia research again provides a revealing example, this time in relation to the meaning of "isolation" (Bedoya et al. 2006). The problem that the authors address, using samples from Oriente and from the capital city of Medellín, is that Antioqueños have high levels of European ancestry as detected by autosomal markers (around 80 percent), but also very high levels of maternal indigenous ancestry, as detected by mtDNA haplogroup frequencies (about 90 percent). The story that authors tell with the data is one of initial mixture between indigenous women and European men. The colonial situation in Antioquia was very similar to what historians and geneticists have argued for many other areas of Latin America (Gonçalves et al. 2007), as the authors recognize (Bedoya et al. 2006: 7234). So far, so normal—and there is no argument here about isolation, because the narrative is all about European immigration.

After independence, however, the data show "strong evidence of postcolonial isolation" (Bedoya et al. 2006: 7236), as immigration declined and the population expanded through sexual reproduction. The evidence for this isolation is provided by the continuity and frequency of common surnames, in the founding Oriente region, and by the fact that sampled individuals from that region who had the five most common surnames also had rather low haplotype diversity. One might expect such low diversity from men with the same surnames sampled in a small population spread over six towns all within fifteen miles of each other.

However, the whole argument about postcolonial isolation adds little to the overall explanation of indigenous ancestry in the mtDNA and European ancestry in the autosomal DNA, as this combination is mainly explained in the article by colonial processes, which are presented precisely as not indicating isolation. The isolation thesis ends up being an argument almost for its own sake, made on the basis of the Oriente samples. At the end of the article, it is linked to broader themes, by contrasting a model of admixture based on postcolonial isolation to one based on the massive nineteenth- and twentieth-century European immigrations that affected countries such as Argentina: it is the postcolonial isolation that has allowed the initial colonial mix to remain relatively stable. (The point is that admixture mapping might need to be tailored to such different populations.) But, by this standard, the whole of Colombia could be taken as postcolonially isolated by comparison with Argentina.

What stands out is the way the image of an isolated population is made specific to Antioquia, as if the region were special in this respect. The geneticists who worked in the lab we studied in Medellín were very clear about the specificity of the paisas: "Antioquia and the paisas are extremely European, you can see that in the genome and in the customs. Also, people were endogamous and they only married among themselves, and that continues to happen today. That's why we are a genetic isolate, see? So I think we'll keep on being very European here, not like in other parts [of the country] in which you can see that they have more of the Amerindian or of the African." The power of the idea of the paisas, as different and distinctive, is such that it subtly shapes the published analysis, pushing the argument about isolation to the fore, when the conclusions of the article do not really require it.

A second telling example of the way underlying categories shape the interpretation of data is the persistence of the indigenous/mestizo conceptual classificatory divide in the research on Mexican genomic diversity. García Deister (chapter 6) found that the basic idea that indigenous and mestizo populations were different and separate underlay the sampling processes, which initially excluded indigenous people and then was obliged to include them. It was clear to INMEGEN researchers which kind of population had to be accessed: indigenous peoples spoke an indigenous language (as did their parents and grandparents) and had indigenous *usos y costumbres* (ways and traditions) and classified themselves as indigenous. The Mexican state marked them out on a map.[14] Researchers from INMEGEN also had to mobilize different academic disciplinary networks—anthropologists and rural doctors—to get to them. The data showed that Zapotec samples looked genomically distinct from the mestizos samples (Silva-Zolezzi et al. 2009). Yet it was also the case that Silva-Zolezzi recognized that people classified as indigenous were, in a certain sense, inevitably mestizo, after more than five centuries of colonial domination and interaction: "In individuals that we regard today as being almost 100 percent indigenous, there must probably exist traces of admixture that we are currently not capable of detecting. Because if you find in 4 out of 30 individuals clear evidence of admixture—even if it is in a very small proportion, but it's there—it is to be expected that all the population to which this individual belongs has a certain trace of mestizaje." Although the data from the samples looked different, we also found that in the lab indigenous individuals who looked genetically distant from the main cluster and were located closer to the mestizo samples tended to be defined as outliers and excluded from the indigenous data sets. The whole point of the Zapotec sample was that it served as a proxy for Amerindian ancestry: it therefore had to be as unmixed as possible. The effect of this was to produce an image of Zapotecs as more genetically distinctive than a different sampling technique would have done. It would have been possible to think of all of the samples as a single category, with greater or lesser percentages of different ancestries; instead "indigenous" as a radically distinct biosocial category in the Mexican national imaginary was reproduced as a category for organizing genetic diversity in the lab.

The case study of the Gaúcho research provides a further example of how data are interpreted by researchers (see chapter 4). The researchers argue that data showing mtDNA with 52 percent Amerindian ancestry in their samples "revealed that the known cultural continuity between pre- and post-Columbian

Pampa populations was also accompanied by an extraordinary genetic continuity at the mtDNA level" (Marrero et al. 2007: 168, 169). Bortolini's earlier work showed that Brazilian black populations had proportions of African ancestry in their mtDNA that were higher than the Gaúchos' 52 percent Amerindian ancestry.[15] Yet the conclusion drawn from these data is that a "model based mainly on admixture provides a reasonable explanation for the genetic structure of these African-derived populations" (Bortolini et al. 1999: 559, 557). Links to Africa are not denied, but neither are they privileged in the way the Amerindian—specifically Charrua—maternal ancestry is highlighted for the Gaúchos. This is all the more striking when one notes that, in a few of the populations Bortolini and her team researched, African ancestry was very high when looking at both mtDNA (58–82 percent) and the Y chromosome (55–96 percent). That is, in some cases (including black individuals in urban Ribeirão Preto), African ancestry predominated in the paternal and maternal lineages and continuity with Africa, rather than admixture, could have been emphasized. Yet the only remarks made on these cases state simply, "No evidence for asymmetrical matings in relation to sex and ethnic groups could be found"; that is, there is no evidence in these cases of the "most consistent finding," which is "the introduction of European genes through males" (Bortolini et al. 1999: 560). The important genetic contributions of African male and female ancestors goes unremarked. The comparison between the Gaúcho project and the research with black populations suggests that other emphases were possible in the interpretation of these data.

A final example of how genetic data get drawn into different arguments comes from a comparison of earlier and later work by Sérgio Pena and some of his colleagues. In a 2004 publication, data are used to demonstrate that many Brazilians (86 percent) have more than 10 percent African ancestry in their autosomal DNA, while also showing that many people classed as preto and pardo have substantial European ancestry. As noted earlier, this is part of an argument that affirmative action programs that target black people cannot be based on genetic evidence. The researchers also use data from a single rural community to indicate that pretos and pardos have relatively similar amounts of African ancestry and could be considered as a single category (as, the authors note, the black movement in Brazil seeks to do; Pena and Bortolini 2004: 42).

In a 2009 article, which focuses more on how everyday social classifications of race and color interact with perceptions of genetic ancestry, a rather different argument is being made. While many of the same points emerge, the authors focus now on how pardos share a great deal of ancestry with brancos. They argue against a tendency to class pardos and pretos together on the basis of shared socioeconomic traits—potentially forming the basis for health policies for the "black population"—as the data "point to closer biological proximity of browns and whites" (R. V. Santos et al. 2009: 800). This comparison illustrates how different research projects using different samples and genetic markers can result in different images of the Brazilian population if findings are generalized from the sampled population to that of the country as a whole.[16] In this case, the result that established the proximity of pardo and branco is foregrounded, rather than the association of pardo with preto. It resonates more closely with the cautions being outlined in relation to racialized public health policies. This second approach, in turn, is consolidated—and gains additional nuances—in a 2011 article, in which Pena and colleagues focus on how "non-White individuals" in major areas of Brazil have "predominantly European ancestry." They emphasize the "considerable sociological relevance for Brazil" of these results, arguing in particular that they do not provide support for the idea that more than half of Brazil's population is nonwhite, which underlies affirmative action policies (Pena et al. 2011: 7).

Conclusion

The focus of this chapter has been quite precise: we have explored ways in which natural-social categories of nation, region, race, and ethnicity—all filtered through and coloring the overarching concepts of population and ancestry—entered into the practice of particular sets of scientists engaged in specific projects. It has not been our intention to debunk the results produced by these scientists, or to reveal them as false in some way. This is not the agenda of social studies of science in general, despite some polarized debates on the matter (see Stengers 2007). The idea is not to show how extraneous social factors enter the laboratory and contaminate the results, because the social is not a separate sphere from the scientific (Latour 2005; see also introduction, this volume). Instead we have tried to reveal how some of the categories with which scientists work are natural-cultural assemblages and how scientists participate in the process of assembling them: "Amerindian ancestry," for example, emerges as an object out of the DNA of specific contemporary indigenous groups, sampled by virtue of their indigenous identity and the access provided by specific intermediaries (anthropologists, rural doctors) who work with a social and demographic definition of these groups.

In particular, our aim has been to show how particular sets of facts are produced and gain currency, rather than other sets of facts, which might have been, and sometimes are, produced. The scope of our project, covering different labs and different countries, gave us a good insight into this diversity, while also linking local lab activities to wider networks, and enables us to make a contribution to understanding how these processes of assembling facts operate. How sets of facts emerge and become stabilized is shaped by a multistage process of defining agendas, choosing samples, and interpreting and presenting data, all of which occurs in an institutional setting with national and transnational dimensions. In this process natural-cultural categories and objects, such as population, region, nation, and biogeographical ancestry, emerge, not *ab novo*, but building on existing forms and lending them new shapes. In general, the use of certain categories can be presented as a matter of technical, scientific criteria: certain populations are chosen or sampled in particular ways because this obeys the scientific objectives of the project. Yet the use of certain categories influences the results of research—for example, the use of departments as sampling units in Colombia; or the use of rooted populations in a number of different projects. In addition, the choice of categories inevitably creates consequences in terms of focus—for example, that mestizos and mixture are a central focus of many projects, with black and indigenous people often seen as separate from the rest of the mixed nation, or as explicable in terms of mixture, or sometimes as anachronistic and representative of ancestral inputs. All of these categories or focuses could have been otherwise.

The strong impression that emerges from this chapter is of a scientific practice in which, through these processes of assembling facts in particular ways, there is a strong tendency to implicitly reproduce familiar categories of nation, ethnicity, and race (in the sense that ancestral populations look like familiar racial categories to the nonexpert eye). Certain ideas about gender relations are also reinscribed, with dominant European males introducing genes into populations through Amerindian and African women and thus creating national populations. This is a way of narrating particular genetic facts that foregrounds early colonial encounters between Europeans, Amerindians, and Africans and backgrounds subsequent interactions between mestizos, thus also privileging the sexual agency of European men and making African and indigenous men (and European women) less visible (see also conclusion, this volume).

It is very important to understand that the geneticists themselves strongly contest the idea of race and would distance themselves from the idea that they might be reproducing this concept in some way. Our argument is that (a) the use of the concept of European, Amerindian, and African ancestries, and (b) the sampling of populations identified as African-derived or sometimes black, Amerindian/indigenous, mestizo, and, in the case of Brazil, white tend overall to reinforce in the minds of nonexperts the idea of *las tres razas*.

However, it is also necessary to grasp that this is more than a question of pure scientific categories becoming misunderstood when they enter the public sphere. The racialized concepts in question have been in a long-term process of circulation in which boundaries between science and society are not clearly demarcated: these concepts were already natural-social assemblages in which former generations of scientists had an important role to play.

This argument does not mean that genomics did not also change these concepts in the process of enunciation. If population genetics tends to facilitate the reinscription of racialized concepts, this does not mean genetics simply reproduces older versions of the same concepts. The language of ancestry, allelic frequencies, and genetic populations is not the equivalent of the language of raciological science: its character is not typological in the same way, nor does it link culture to biology, nor assign hierarchical value.

Nevertheless, the very process of providing a biologizing and geneticizing idiom in which to describe and imagine populations creates certain effects. Previous research suggests that the effect, at a broader level in society, may not be a simple one of unidirectional geneticization in which the public at large begins to think of identity and belonging predominantly in a genetic idiom (Condit 1999; Condit et al. 2004; Nelson 2008b; Wade 2002b, 2007a). But this kind of genomic research clearly opens the possibility of understanding socially defined populations in genetic terms, even if what people do with this possibility is very varied. More specifically, the genomic research we observed creates specific effects in its practice. The possibility of biologizing and geneticizing has been open for some time, but recent genomic research into human diversity, in which race is refigured as biogeographical ancestry and attached to regions, nations, and genders, creates specific possibilities related to the individualization of ancestry, the precision of calculating ancestry, the revealing of hidden and very particular matri- and patrilineages, and the abstraction of ancestry. (These themes are developed in the conclusion.) In the process, the natural-cultural entanglements in assemblages of ancestry, region, nation, and, implicitly, race, become ever more complex and highlight the hybridizations that Latour (1993) saw as always underlying the purifications that tried to separate nature, the realm of science, from culture, the realm of society.

Notes

1   One publication on "Latin American mestizos" said the research had "a number of implications for the design of association studies [associating disorders with genetic variants] in population from the region" (Wang et al. 2008: 1). A Colombian article suggested that the population of "Antioquia (Colombia) is potentially useful for the genetic mapping of complex traits" (Carvajal-Carmona et al. 2000: 1287).

2   See "Censos y Conteos," Instituto Nacional de Estadística y Geografía, http://www .inegi.org.mx/est/contenidos/Proyectos/ccpv/default.aspx. The 1930 and 1940 censuses counted monolingual and bilingual speakers separately: they were roughly the same number.

3   For a more recent statement suggesting that race may not be simply a social construction, see Salzano (1997).

4   We encountered the category "Caucasian mestizo" only in publications by Colombian geneticists. It reflects the tendency in Colombia to assimilate mestizos and whites.

5   See also "Qué es la Expedición Humana?," http://www.javeriana.edu.co/Humana/ humana.html.

6   Compare similar accusations made about the Human Genome Diversity Project (Reardon 2005). One result of the controversy around the Expedición Humana is that sampling indigenous populations in Colombia has become increasingly difficult, reinforcing the focus on mestizos. In Brazil too, ethical controls have made it increasingly difficult to sample indigenous people, which has reinforced both the sense of them as a separate category and a focus on nonindigenous people (Santos 2006).

7   In characterizing the ancestry of the mineiro, the geneticist referred to the analysis of haplogroups done on the mtDNA of white male individuals featured in the *Retrato Molecular* (Pena et al. 2000: 22).

8   The final set of terms was: "negro," "mulato," "indígena," "moreno" (brown), "cobrizo" (copper), "mestizo," "blanco," "pardo," "Europeo," "otro" (other).

9   The researcher was referring to the widespread perception in Brazil and other areas of Latin America that black people often try to avoid identifying or being identified as black—for example, in the Brazilian census—due to the low social status associated with blackness (Telles 2004; Wade 1993).

10   See also the map in Barragán Duarte (2007). This too used departments as sampling units.

11   See for example the regions outlined by the official Instituto Geográfico Agustín Codazzi, http://geoportal.igac.gov.co/mapas_de_colombia/IGAC/Tematicos/34813 .jpg. These can be found in numerous regional maps of Colombia.

12   For example, it is used in the People of the British Isles project, in which volunteer DNA donors have to have "parents and grandparents [who] were born in the locality" (see "Information Sheet," Ver. 1.5, October 2009, http://www .peopleofthebritishisles.org/vi/).

13 In Brazil, the reference population that is used at present in such cases is still the one provided by the FBI (Kent, field notes).

14 See for example the website of the Comisión Nacional para el Desarrollo de los Pueblos Indígenas, http://www.cdi.gob.mx.

15 These proportions varied from 90 percent in urban Salvador to 58 percent in the village of Paredão in southern Brazil (Bortolini et al. 1999).

16 While the 2004 publication was based on ten markers suitable for differentiating between African and European ancestry, the 2011 publication was based on analysis with forty INDELS that were designed to differentiate between African, European, and Amerindian ancestry.

Race, Multiculturalism, and Genomics
in Latin America

Peter Wade

This conclusion tackles the broad questions raised by our research, as reported in the previous chapters. Bearing in mind that our project focused on a handful of laboratories and projects—and that even these presented a good deal of diversity—in what follows I seek to draw out general conclusions that emerge from our research. Are categories of race, ethnicity, nation, region, and gender being revived, reproduced, refigured, deconstructed, or abandoned; are they being regrounded in biology via genetics? What specific reconfigurations do different types of genomic work produce? What does our Latin American material have to say to the broader debates on genomics and race (and ethnicity, nation, etc.)? Is something specific lent by the general context of *mestizaje* as an ideology of national identity or are the processes at work in the genomics labs and beyond them basically the same as those in North America, Europe, and elsewhere? How does the knowledge of human diversity produced by this kind of genomic research relate to state policies of multiculturalist governance and to regimes of citizenship and power in these Latin American nations, bearing in mind postcolonial preoccupations in these countries with global, and especially hemispheric, positionings? What can the process of comparison between Brazil, Colombia, and Mexico tell us and, more broadly, what work is being done or implied by the very process of national comparisons? Does it make sense to treat Latin America as some kind of unit in a global comparison? The aim in what follows is to draw on the detailed material and arguments in the book to provide some answers—from the point of view of our particular project and the labs we studied—to these questions.

Common and Enduring Concerns

As outlined in the introduction and part I, the concerns that genomic research addresses are long-standing ones. First, in all three nations, elites have long pondered the implications of being the product of mixture between Europeans, Africans, and Amerindians, a trait that has been seen as both problem ("racial degeneration") and opportunity (mestizaje as national identity). Second, the three countries all have a history of concerns among scientists, medical doctors, reformers, and administrators with questions of race, mixture, and health, typically voiced in the early decades of the 1900s through a eugenic discourse, which addressed fitness in both physical and moral aspects. Third, an interest from scientists in the physical anthropology of human diversity in the nation is also long-standing. Early studies in the late 1800s and early 1900s tended to focus on indigenous groups (and black populations in Brazil), but the overall project soon included an exploration of the biological diversity of the nation, with an explicit focus on the mestizo emerging from 1911 in Brazil and a little later in Mexico and Colombia. Fourth, in all three countries, ideologies of mestizaje or *mestiçagem* as the core identity of the nation, present and future, have been strongly tempered by challenges from black and indigenous social movements and, from about 1990, by official multiculturalism.

Through all this, as noted in the introduction, the term and concept of race have been an absent presence in these countries. They were used extensively in the early decades of the 1900s, but as early as the 1920s the term at least began to be avoided. Race was then often denied as having much significance in social life, while everyday reference to *negros, indios, blancos, mestizos*, and so on was common and, in the Brazilian census, was enshrined as color (*preto, pardo, blanco*, etc.). Meanwhile, some (but not all) black and indigenous people—and some white people—protested that racism was an issue. With multiculturalism has come some greater openness about racism and thus about racialized difference, but many people still see race as a foreign optic through which to view their societies. The term and the concept fade in and out of focus and are simultaneously present and absent.

These are some of the long-standing concerns against which we have to understand the recent science research on genomics, mestizos, ancestry, diversity, and health. The questions with which geneticists are grappling—population history as revealed in DNA markers, the identification of genetic variants associated with complex disorders, the genetic diversity of national populations, and so on—are a continuation of older preoccupations with race, mixture, diversity within the nation, health, education, and nation building. The question is, how does genomics, as a particular set of technologies and knowledges, inflect these concerns?

Genomics in Society: Changing Regimes

Within the frame of the role played by new genetics in society (see introduction), there are specific debates about what is happening to concepts such as race, ethnicity, and nation with the advent of genomics. A key element in these discussions is whether race (with rather less attention to other related categories) is being revived, reproduced, transformed, taken apart, or banished; and whether it is being regrounded in biology via genetics; and how nonexpert publics then react to this.[1] Many studies indicate that race does not disappear, although it may be refigured or appear in other guises—perhaps via genome geography, that is, the reference to geographically defined populations that are more or less easily assimilable (depending on who is doing the assimilating) to commonsense notions of races (Fujimura and Rajagopalan 2011). Marshaling recent literature on race, health, and genetics, Abu El-Haj (2007) sums up how race has moved away from the typological categorizations of classic racial science toward a more probabilistic, calculative mode, based on DNA rather than phenotype. Recent observers outline a regime in which the risks of the predisposition of a racialized individual to a given disorder are assessed, thus locating the individual firmly in neoliberal regimes of self-monitoring citizens, who must take responsibility for their own welfare, as well as linking race to regimes of capital accumulation, based on selling race-specific medicine (Kahn 2005, 2008; Rose 2007; Rose and Novas 2005).

This idea of the transfiguration of race into new idioms and processes is an important one. Rather than only asking how familiar categories of race (and ethnicity and nation) reappear in genomic practices—although this is a necessary question—this approach also asks how genomics casts race in new forms and inquires into the effects such forms have as they circulate in society. It delves into the practices of seeing (visibilization) and saying (inscription, enunciation) characteristic of genomics; it explores the regimes of truth established in genomic science; it looks at how genomics assembles specific objects (DNA, AIMs, genetic ancestry, etc.), which are new and also subject to change. Here it is also important to distinguish genetics—which begins in the early 1900s and focuses on single genes and associated traits—from genomics—which dates from around the 1980s and, enabled by advances in DNA sequencing, studies whole genomes and the gene-environment interactions that underlie complex disorders such as diabetes. One question is what reconfigurations are introduced by the different types of genomic research we were studying.

In framing these questions, it is helpful to think in terms of a general shift that Deleuze, in dialogue with Foucault, characterizes as a move away from regimes of discipline toward regimes of control (Deleuze 1992). Rather than governing people through enclosure into, for example, prisons, insane asylums, schools, barracks, and factories, which all involve regimentation and standardization, control is achieved through constant modulation and reflexive adjustment. People become data, which are subject to constant change, adapting to varying circumstances. This is reflected in the numbers regime of population metrics, which, rather than relying on occasional censuses—which pin people down in one place at one time (Rusnock 1997)—continuously collect data on transactions and movements and thus constitute a population as "a complex assemblage of transactions, movements and conduct that modulates, changes and transmutes and must be tracked and measured on an ongoing basis" (Ruppert 2009: 13). The key point here is the emphasis on movement, fluidity, transformation, and lack of fixity—processes for which genetics and digital computing have been key factors, according to Deleuze (2006: 109). As Rose and Novas (2005: 442) say, "Biology is no longer blind destiny. . . . It is knowable, mutable, improvable, eminently manipulable." The question is: if genomic science practices entrain this kind of fluidity, movement, and constant modulation, is this reflected in the way race-like concepts appear in the practices of Latin American genomics labs? The answer is, perhaps predictably, yes and no.

Fixing and Transfiguring Race, Region, and Nation through Genomic Research

In an important sense, the genomic research we explored involved a good deal of the fixing, in very precise and numerical ways, of populations, regions, nations, and racialized ancestries. As outlined in the introduction, the specification of biogeographical ancestry (African, European, Amerindian) can be considered racialized—despite the rejection of race as a valid category by the geneticists we worked with—because it relies on familiar, race-like categories, which are refigured in an idiom of genetic ancestry. Whereas older genetic studies often talked unproblematically about racial mixture, more recent ones avoid race and talk instead about ancestry or the frequency with which, say, Amerindian haplogroups appear in the DNA of a sample population. Thus race-like ideas are being refigured through concepts of biogeographical ancestry (Bliss 2009a; Fullwiley 2007a, 2008; Reardon 2005; Roberts 2010). Such racialized categories are far removed from the raciological science of the early twentieth century—in that they are probabilistic and much less determinist—but they are racialized nonetheless.

As the chapters in part II of this book show, many labs in Brazil, Colombia, and Mexico produced tables and charts with statistical breakdowns of the exact percentage contributions of African, Amerindian, and European ancestry to the DNA of particular populations.[2] Some publications used simple tables to

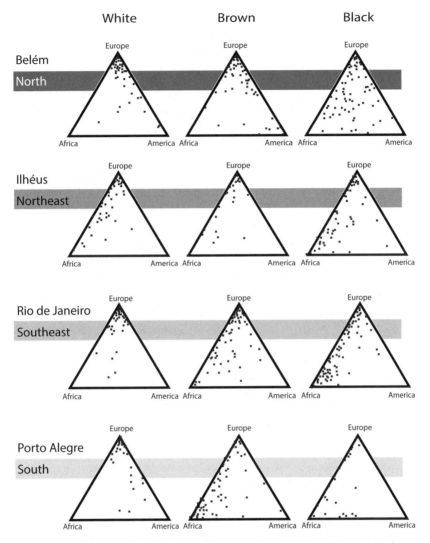

Figure C.1. "Triangular plots of the genomic proportions of African, European and Amerindian ancestry in three self-reported color groups of 934 Brazilian individuals from four different regions of the country, self-categorized White, Brown and Black individuals" (Pena et al. 2011).

represent ancestral contributions, others used pie charts and bar charts, while some used devices that placed individual people on a relational grid, indicating their position in relation to each other and to reference points, taken as parental populations of Africans, Europeans, and Amerindians or proxies for them (see figures C.1–C.3 and table C.1).[3] These inscriptions and visualizations

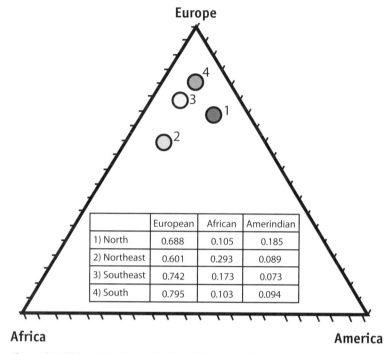

Europe

	European	African	Amerindian
1) North	0.688	0.105	0.185
2) Northeast	0.601	0.293	0.089
3) Southeast	0.742	0.173	0.073
4) South	0.795	0.103	0.094

Africa   **America**

Figure C.2. "Triangular plot and table of the genomic proportions of African, European and Amerindian ancestry in four different regions of Brazil, independent of color category. Each point represents a separate region, as follows (1) North (Pará), (2) Northeast (Bahia), (3) Southeast (Rio de Janeiro) and (4) South (Rio Grande do Sul)" (Pena et al. 2011).

are important because tables and charts, as objects, condense so much data into forms that are easily made mobile and reproduced (Latour 1990). Thus percentages are routinely reproduced in newspaper accounts of genomic projects;[4] and a map of Mexico showing different proportions of continental ancestries by region tops the Facebook page of the CANDELA genomic diversity project in Mexico.[5]

The use of genetic markers to calculate ancestral percentages contributing to an admixed population has been common practice for a long time. Researchers used classical genetic markers (e.g., traits found in blood groups and serum proteins linked to specific alleles) to make calculations of what was often called "racial mixture" (Franco, Weimer, and Salzano 1982; Ottensooser 1944, 1962; Salzano and Bortolini 2002). Genomic science thus reproduces previous techniques in some respects (see Marks 1996). However, new technologies allow much greater numbers of genetic markers to be used, linked to

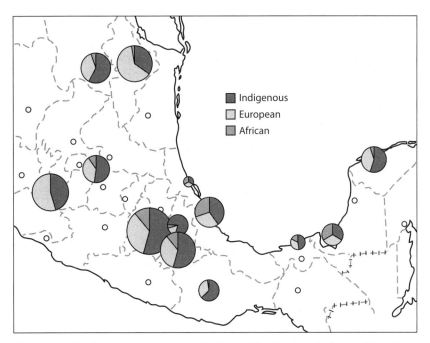

Figure C.3. Map showing ancestral contributions of Mexican populations, with estimates from a trihybrid model based on autosomal markers (courtesy of Víctor Acuña Alonzo).

more comprehensive databases of reference samples; they also allow analysis of different parts of the genome, such as mitochondrial DNA, Y chromosome DNA, and nuclear DNA, which give data on different aspects of ancestry. In the following sections, I explore some features that stand out in these genomic inscriptional devices and visualizations, which are different from earlier regimes, while showing some important continuities.

Fixing Foundational Categories

First, these inscriptions tend to reproduce certain foundational categories and geographies. Latin American populations are being fixed in relation to constantly reiterated African, European, and Amerindian genetic reference points, which resonate easily with familiar concepts of *las tres razas* (the three races), which are popularly seen as the foundation of the nation.

This is evident in the plots shown by Silva-Zolezzi et al. (2009), which graphically map genetic distances between individual people and populations. The plots show clusters with geographical labels such as CEU (the HapMap label for Utah residents with northern European ancestry) and YRI (the HapMap

Table C.1. Example of Table Showing Ancestral Proportions, adapted from Pena
et al. (2009: table 1)

Average Values of the Genomic Proportions of Amerindian, European and African
Ancestry of Individuals from the CEPH-HGDP Panel and from the Brazilian Population

		Ancestry		
	Number	Amerindian	European	African
CEPH-HGDP population				
Europeans	161	0.028	0.946	0.026
Amerindians	105	0.948	0.037	0.014
Sub-Saharan Africans	126	0.011	0.023	0.965
Brazilian Whites				
Southeast (Minas Gerais)	142	0.131	0.708	0.161
North	45	0.121	0.742	0.137
Northeast	49	0.147	0.711	0.142
South	36	0.092	0.819	0.089
Southeast (São Paulo)	88	0.105	0.779	0.116
Brazilian Blacks				
Southeast (São Paulo)	100	0.132	0.371	0.497

All individuals were typed for all 40 insertion-deletion polymorphisms and analyzed with
the Structure program using the CEPH-HGDP populations as references.

label for Yoruba from Ibadan). This can be accepted at face value because the
HapMap samples that these labels refer to and the software used to produce
the plots are standard in the field (Bliss 2009a; Fujimura and Rajagopalan 2011;
Reardon 2008). They both carry with them a baggage of biogeographical populations, which emerges as normal and unremarkable in these plots of Mexican samples. The plots are able to fix the Mexican mestizo populations—by
specifying their location and enumerating their ancestries—because of the
nature of the reference samples and the software used to analyze the data.

This underlying racialized structuring of Latin American populations,
made visible in the tables and charts that constantly placed them in relation
to the three parental ancestries, was noted in chapter 7. There we saw how
foundational national dualisms that opposed indígena to mestizo and African-
derived to mestizo were often used as organizational devices in genomic research—a trait that is shared with earlier genetic research on admixture. This
occurred despite the possibility of seeing both indigenous and African-derived
populations as, to a greater or lesser extent, genetically mestizos. On the one

hand, then, researchers showed that social categories of identity could not be defined in terms of genetic ancestry, as the categories overlapped considerably with each other; on the other hand, familiar categories of identity were reiterated and even apparently given genetic meanings, in terms of greater or less degrees of admixture.

Linked closely to this structuring of diversity by foundational, race-like categories, we saw that commonsense geographies of the nation were commonly reproduced in the inscriptions used in presenting genomic data. The populations whose ancestries were under study were generally defined as relating to specific regions or territorial administrative units (departments, states, major regions—and ultimately the nation itself) and were shown in tables and, often, maps, which reproduced standard and usually official geographies of the nation. In addition, as we saw in chapter 7, many sampling procedures selected individuals who had parents and grandparents born in the spatial or population unit that was being sampled, thus fixing the population as ancestrally rooted in time and space.

Fixing through Precision and Triangulation

The second feature of these ways of enunciating and visualizing populations is the mathematical precision of the admixture estimates (albeit they were often hedged with standard margins of error). Mathematical precision in assessing race is not new: anthropometrics used exact physical measurements of anatomy, skin color, eye color, and facial features (Sá et al. 2008).[6] The key difference is that these older methods were based on phenotype and used to assign individuals to a racial category, defined in biological terms. Later studies used classical genetic markers to calculate ancestral proportions in percentages and assess degrees of race mixture in populations. Now, a wider range of DNA markers is used to refine with ever-greater apparent precision degrees of ancestry for both populations and individuals, even if, at the individual level, the precision is not as great as it appears. The new technologies allow populations to be fractionated and individuals to be fixed and differentiated from each other with apparent exactitude—but always in relation to parental reference populations. If it is the case that "when qualities are successfully quantified, they also become objectified" (Wise 1997a: 7), then ancestry as a quality becomes a fixed object, rendered in numbers that make each individual and population comparable. The precise metrifications generated suggest a unity insofar as all samples are in principle reducible to the same calculus— precision is an "agent of unity and a product of agreement" (Wise 1997b)— which fixes ancestry in relation to the same three parental populations.

Related to this question of precise individuation is the particular nature of certain common techniques for exploring ancestry, which emerged with DNA sequencing and which were not available to the racial mixture studies based on classical genetic markers—I refer to the analysis of mtDNA and Y chromosome DNA. Whereas autosomal DNA tests estimate ancestral contributions by identifying markers in all of an individual's DNA, inherited from many thousands of ancestors, mtDNA and Y chromosome tests analyze DNA that has been passed in a single unbroken line of descent, either maternal or paternal.

With the addition of mtDNA and Y chromosome analysis, the ancestry of an individual or a population can be assessed three different ways: a person or population may have certain proportions of African, Amerindian, and European ancestry in the autosomal DNA, but be connected in different ways to these parental populations in terms of matriline and patriline. Individuals and populations can be fixed in a three-way triangulation.

This is all part of a gradual refiguration of race: precisely specified percentages of ancestry or frequency of genetic variants give an image of belonging more or less, finely graded by degree of admixture. The possibility, long present in Latin American mestizaje, of thinking in terms of multiple ancestral belongings is vastly increased and lends a new precision and fixedness. The image of precision and triangulated fixing derives from new DNA sequencing technologies, but these still relate to parental populations as reference points. And these ancestral categories are very familiar ones, even though they appear in a new guise as sets of DNA markers—or rather the frequency of occurrence of those markers—which stand for contemporary populations of Europeans, Africans, and Amerindians, which in turn stand for parental populations. As a result, familiar ancestral categories increase their salience, while remaining very easily assimilable to concepts of las tres razas, which are popularly seen as having formed Latin American populations and nations.

Fixing through Impersonal Abstraction

A third notable feature of the inscriptional devices used in this genomic science is that ancestry, while being individualized, is depersonalized and made abstract. Previously, racialized ancestry might be guessed at through visible phenotype. Ancestry could also be identified through personal family genealogy (subject to being manipulated and obscured), which connected a person to other persons through traceable kinship links. Ancestry at a population level could be inferred through other cues assumed to be relevant, such as region of origin or perhaps social class. Older genetic techniques could specify percentages of ancestry—for populations, rather than individuals—and this began to lend an aura of greater reliability than the inferences based on appearance, family history, and region of origin.

The analysis of mtDNA and Y chromosome DNA reveals aspects of population ancestry inaccessible to earlier genetic techniques. These invisible, inaccessible aspects of ancestry relate to the individual, but are also curiously impersonal. A person, or indeed a population, can be linked to unknown ancestors, who might well be totally inaccessible to historical genealogical tracing.[7] The link is typically made in abstract terms of haplogroup membership or percentage ancestry. One can say for an individual that mtDNA analysis reveals that he or she belongs to a given haplogroup of people, who share a mutation inherited from a distant common ancestor, more or less locatable in time and space. One can say for a population that mtDNA analysis shows that 33 percent of haplogroups found among its members are Amerindian (in the sense that these haplogroups are frequent among Amerindian populations). In either case, this reveals an extremely specific aspect of the total ancestry: for each individual sampled, it selects one unilineal genealogical line out of all myriad possible ancestors (M'charek 2005a: ch. 5; TallBear 2007).

The significance of the revelation made by this kind of analysis—and revelation it is often said to be for individuals (Gaspar Neto and Santos 2011; see also R. V. Santos et al. 2009)—is subject to all kinds of mediation through individual histories and perceptions (Nelson 2008a), but the point is that it appears as a hidden reality that has been made visible and factual. The information may be denied, ignored, or forgotten, but it has a new level of perceived reliability—even if, at an individual level, there are wide margins of error (Bolnick 2008). It appears that DNA reveals a deeper level of personal and populational truth, but in an impersonal way.

Previously, individual ancestry was embodied in family relations, kinship, sexual reproduction, and "blood"; ancestral proportions were subject to the genealogical calculus of cognatic kinship (half, quarter, eighth, etc.). With DNA analysis, personal ancestry becomes distant from the domain of kinship and family; the relation is no longer between people, but between numbers. In a sense, molecular technologies generate a version of Anderson's "homogeneous, empty time" in which unconnected citizens in a nation can imagine themselves simultaneously sharing an activity—such as reading the daily paper—that creates an identification (Anderson 1983). In the same way, learning that one shares specific, but normally invisible, genetic traits with unknown others has the potential to create imagined genetic communities—although these would not necessarily be national in scope. Such knowledge can act as the scientific proof of imaginary consanguinity.[8]

Genomic Grids of Intelligibility

The foregoing points about foundational categories, precision, and abstraction all indicate that certain—but not the only—aspects of the genomic research we studied tend to impose rather inflexible grids of intelligibility, in which individuals and populations are defined, rooted, and fixed in ways that rely on stable categories of race, ethnicity, and nation—ones that are rather well known and draw some stability from this familiarity. As the following section argues, the same science contains other possibilities, which tend to unfix race, but it is important to simultaneously grasp the potential for fixing. Far from simply creating or participating in a regime of constant modulation and change, there are strong indications of reification and stabilization. In doing this, genomics makes recourse to familiar categories and concepts, but also achieves this fixing in new ways. It is not a question of the simple reinscription of old categories. Nor is it even just a matter of the provision of data that could be "misinterpreted as indications of 'real' racial divisions" (Shriver and Kittles 2008: 209), although this remains a distinct possibility. Rather, it is the refiguring of race-like categories and ideas in a specifically genomic idiom that entrains multiple reifications, which all reinforce each other—the precision of finely graded diversity and the intense fractionation of population in relation to ancestral categories; locating precisely differentiated ancestries onto geographies of the nation and the globe; construing individual ancestry as abstract genetic proportions, as distinct from personal family genealogy; focusing on unilinear connections that reach back into the remote past and are extremely selective among all possible ancestors.

Unfixing Race?

The fixing of categories of race, ethnicity, region, and nation is only part of the story. The very same techniques that fix—the precise numbers, the fractionation, the visibilization of hidden genetic markers as abstract ancestry, and so on—simultaneously create other effects, which tend in a different direction. Referring to the genetic code, Deleuze remarks that the classic nineteenth-century social formations charted by Foucault worked between a conception of finitude (the realm of humans) and infinity (the transcendent realm of God). With explorations into molecular biology and silicon-based forms of information processing, the forces at play now involve "an unlimited finity," evoking a situation in which "a finite number of components yields a practically unlimited diversity of combinations," as borne out "by the foldings proper to the chains of the genetic code" (Deleuze 2006: 109).

The Unlimited Finity of Ancestry

Although Deleuze is talking at an abstract level, unlimited finity is precisely what we can observe in the numerical precisions of ancestral components, defined down to the individual level, and in tables of allelic frequencies. Ancestral proportions are fixed in one sense, yet there is always the sense that a different combination could be possible and that the number of combinations is practically unlimited. This is partly a matter of the numbers being qualified by margins of error, but it is more an effect of the technologies themselves and of being precise in relation to populations.

First, despite the standardization of methods, different figures will be produced by different samples and different sets of genetic markers; there is a lack of finality here, a permanent deferral of the last word, inherent in the methods themselves.[9] Related to this is the multiplication of ancestry by using mitochondrial, Y chromosome, and autosomal DNA. This can be fixed through triangulation, but it also multiplies the very concept of ancestry, giving several ways of perceiving the relation of individuals and populations to the three founding parental populations. As a way of thinking about the individual or the nation, mestizaje has always carried both the image of homogeneous similarity and the suggestion of diversity, multiple origins, and a mosaiclike constitution. Genomics intensifies and reifies this by providing at least three ways of understanding personal or population ancestry.

Second, if the population of Brazilian major region or a Mexican state is presented as having 4.2 percent more African ancestry than the region or state next door, it is already impossible to think of that change occurring exactly at the regional or state border. These are necessarily generalizations that could always be different; and each individual will be some variation on that very precise figure. Sequencing of DNA has intensified and multiplied the differentiations that are produced, as well as making them more precise. In the graphics used by Pena et al. and by Silva-Zolezzi et al., the ancestral makeup of individuals is shown, giving a real sense of that intense heterogeneity and fractionation—that each individual is different. This, of course, is precisely the claim that Sérgio Pena himself makes—that we have to think of the Brazilian nation as 190 million individual citizens (Pena et al. 2009, 2011).

As we have seen, the individual data are nevertheless relentlessly organized into categories (Mexican states, Brazilian regions and census color/race categories, Colombian regions) and are always related to foundational ancestries, but it is possible to imagine that other categories would be possible (Mexican municipalities, noncensus self-classifications), which would break up the data in multiple ways, with multiple estimates of ancestral admixture. When

Paredes et al. (2003) produce a regional map of Colombia, derived from clus-terings calculated on the basis of allelic frequencies, and another researcher later produces a slightly different map, the difference is said to be due to a bigger sample and a different set of markers being used, and the difference is presented as an improvement.[10] Yet this immediately implies that no map will ever be final: there could be an unlimited finity of maps, each slightly different from the others. The nature of these stochastic processes, as they are enabled by the diverse technologies and methodologies of genomics, means that the very precision that statistical analysis allows is always undermined by a sense of imprecision lent by the lack of finitude; nothing is final.

In other words, we again confront a refiguration of race (and ethnicity, re-gion, and nation) through the unlimited finity generated by genomics: some key foundational and reference categories are fixed and reiterated, and pop-ulations and individuals apparently located precisely in relation to them, but the very focus on invisible DNA and the approximation to it through numbers makes those precisions less stable than they appear at first sight. The endless multiplication of possible positionings in relation to the reference points—the fractionation of populations, the differentiation of individuals, the diversifi-cation of ancestries—creates a sense of movement and modulation. The sta-bility of the reference points themselves is also perhaps less than it seems—one could use HapMap data to specify "African ancestry," or one could use HGDP-CEPH data—but this multiplicity is more masked than the multiplicity of the populations with the nation, because it refers to internationally accepted standards, which are used to organize the data.

Uncertain Connections of Ancestry and Disease

A second important dimension in which movement and instability occur around race in genomics relates to the way race, ancestry, and health get con-figured together. When the commentaries on new regimes of risk and control highlight constant movement and flexibility, this is most easily reflected in areas connected to disease (Abu El-Haj 2007; Rose 2007). Here one can appre-ciate how genomics is linked to the specification of racialized risk factors for individuals and populations, which generates uncertainty. A patient may ask, among all the factors that are thought to contribute to diabetes, how import-ant is the fact that I am Mexican? How should I act in the light of knowledge of my genetic ancestry? The answer is that no one really knows for sure (Montoya 2011). In Brazil, Mexico, and Colombia, there have been attempts to correlate percentage ancestries or allelic frequencies with the incidence of certain con-ditions, such as obesity, diabetes, and heart disease (e.g., Acuña-Alonzo et al.

2010; Villalobos-Comparán et al. 2008). In one way, this creates a fixity—a certain percentage of Amerindian ancestry may confer a given risk of diabetes or correlate to cholesterol levels—in another way, the language of risk creates room to maneuver. It is a risk factor (of as yet unknown proportions), not a determination.

Thus, the diabetes prevention campaign of the Mexican state institute for social security, which in 2008–2011 used the slogan "Don't inherit diabetes," focused on healthy lifestyles, with the implication that good habits could help control inherited predispositions. With a complex disorder of this nature, the role played by genetic ancestry is subject to uncertainty: the slogan used the language of inheritance, but this can invoke processes of social inheritance, added to which the campaign's message is about lifestyle, not biology. Precise specification of the proportion of Amerindian ancestry may tell you very little about the causes—much less the treatment—of diabetes in your particular case and the onus is then on you to undertake risk management. A further example is screening for sickle cell anemia in Brazil (Fry 2005b). Health campaigns associate Afro-Brazilians with the condition in some ways, but it is recognized that anyone could be a carrier of the trait, so some state screening programs, based on genetics, screen everyone, not just blacks. The fact that a lot of genomic reference to racialized ancestry occurs through constructs of health and risk means that race becomes an uncertain risk factor, rather than a stable identification.

The Uncertainty of Abstraction and Its Domestication

The abstractness of genomically specified ancestry is another interesting area in which questions arise about the effects of the genomicization of race in Latin America. I argued above that abstraction appeared as a way in which individual racialized ancestry could be fixed as a fact, once hidden, now made visible and even public as a precise calculation of molecular connection. Yet, as with the numerical specification of endless difference, this seems to cut both ways. The very abstraction of the way ancestry is inscribed and visualized, distant from family personal genealogy, makes it highly subject to manipulation. Various studies indicate that people interpret individual ancestry test results in ways that mix *bios* (biology) with "bios" (biographies) in strategic ways (Nelson 2008a: 762; see also R. V. Santos et al. 2009; Wailoo, Nelson, and Lee 2012). There seems to be a powerful need to reconnect the abstraction of genetic ancestry to meaningful tropes of kinship and family, which allow people to relocate the bald fact of a genetic connection, expressed as a percentage or a haplogroup membership, into a social connection that is experienced as more "real." This suggests that abstraction, while it may be fixed in some ways, is in other ways empty and unfixed and has the potential to unsettle familiar identifications. This potential is made evident in the apparent need to domesticate it.

This process of domestication was most evident in relation to the frequently reported genetic evidence of indigenous (and sometimes African) markers in mtDNA and European markers in Y chromosome DNA (Bortolini et al. 2004; Gonçalves et al. 2007). Often this was reported for populations self-classifying or generally seen as white or on the whiter end of the mestizo spectrum. However, this abstract expression of ancestry was often assimilated to a narrative, usually in more popular texts, of indigenous (or African) mothers and European fathers. For Colombia, Emilio Yunis talks in precisely these terms, which are then reproduced by journalists (Yunis Turbay 2009: 117).[11] Mitochondrial DNA comes to stand in for women, even though men also carry it (M'charek 2005a: 140; Nash 2012a). In Brazil, to illustrate their account of these patterns of genetic inheritance, Pena et al. (2000) reproduce a well-known 1895 painting which, via a family scene of grandmother, parents (white man, *mulata* woman), and baby, dramatizes the process (see also chapter 1, this volume).

The genetic lineages established by DNA analysis do not concern the individual; they are abstract lines of inheritance of DNA (M'charek 2005a: 130). Yet talk of indigenous mothers and European fathers relocates the genetic data into embodied, racialized people. This can be seen as a domestication and familiarization of genomic abstractions, which as simple numbers are empty signifiers that need to be reconnected to, and rerooted in, social meanings of family and kinship (Nash 2012a: 8). Something that appeared to be fixed—a statistical enunciation of a molecular connection to indigenous ancestry—turned out, in its abstraction, to be rather destabilizing. What did this connection really mean? But it could be domesticated via familiar stories of European dominant masculinity and receptive, nurturing indigenous and African femininity.

Renaturalizing and Refiguring Race

The balance between these processes of fixing and unfixing cannot be pinned down in a general way: there are multiple possibilities. In this sense, genomics multiplies the options for thinking about race, ethnicity, region, and nation. It participates in the underlying tension between ideas of unity and homogeneity and ideas of diversity and heterogeneity, a tension that is permanently scaled and rescaled. The sharing can be understood to occur from the level of humanity (99.9 percent of all DNA is shared) to the level of, say, a Brazilian region (the

Gaúchos share a genetic continuity with the extinct indigenous Charrua people). The diversity can be envisaged from the individual (each person is unique) to the global (humans are a species among other species). Each scalar level of sharing is also a level of diversity. Genomics provides new ways of reckoning what is shared and what is distinctive.

There is no doubt, however, that genomics creates a new language of biology—a molecular language of invisible traces, grasped statistically—for imagining sameness and difference, and this has particular effects in Latin America. It is well known that in Latin America race is a very culturalized construct (see introduction). The difference between a mestizo and an indigenous person has often been shown to be a cultural one, based on language, dress, place of residence, and so on, rather than a biological one, based on appearance or ideas about parentage. This contrast has been overdrawn at times, with too easy a divide between culture and biology, too little attention to the continuing importance attached to both racialized phenotype and parentage in Latin America, and a tendency to forget that race has always and everywhere been a natural-cultural construct (Gotkowitz 2011b; Nelson 1999; Orlove 1998; Wade 2002b, 2010; Weismantel 2001). Yet there is an important sense in which popular understandings of race in Latin America are potentially being given new and more clearly biological meanings through genomics (Hartigan 2013a). This is not a return to the racial biology of the late nineteenth century (although it responds to some of the same enduring concerns outlined at the beginning of this conclusion). It is a genomic biology that refigures race in new ways. It is manifest in the possibility of genetically placing populations and individuals in relation to African, European, and Amerindian reference points and in abstractly specifying ancestral components at the individual level in terms of percentages and allelic frequencies; in the possibility of imagining a link to an ancient indigenous mother or a long-lost European father, or of sharing an invisible indigenous heritage with many other people who are connected to you via a common ancestor; in the possibility of representing African-derived populations, indigenous populations, and triethnic mestizos as all apparently genetically distinct from each other.

Many observers have argued, for other areas of the world, that race reappears and is being reified in genomic science. The Latin American material contributes to these debates by providing the context of ideological mestizaje in which we can observe with unusual clarity the new possibilities for the refigured biologization of race that genomics affords—in its particular molecular idiom and with all the complexities and unsettlings that this involves. In Latin America, ideas about race have not been without a discourse of nature and biology, but these have been less prominent than in other regions, such as North America or Europe, where more importance has been attached to a combination of inherited phenotype and genealogy in making racialized identifications. Latin America brings into clear relief the biologizing or molecularizing idiom of genomics—abstracted into statistical data—especially the constant reiteration of tripartite mixture and the discovery of the genetic traces of this in virtually all populations (with the partial exception of some indigenous groups, who can serve as a proxy for a notionally unmixed ancestral parental population).

Genomic Research and Multiculturalism

As outlined in the introduction, there have been important shifts in Brazil, Colombia, and Mexico toward an official multiculturalism, which, whatever the debates about its true effects, have certainly focused a good deal of public attention onto Afrodescendant and indigenous minorities and created debate about the shape of the nation.

In this context, it is interesting to explore the social-political-scientific space in which the kind of genomic research in question—and especially the mestizo as a genomic object—can circulate and work effectively and also the kind of impacts they may have. We have seen that genomic research looks at all kinds of populations—including black and indigenous people—but also that the mestizo and above all mestizaje occupy a place of great interest. As the chapters in part I of this book demonstrate, mixture and the mestizo are objects of long-standing interest in networks that include scientists, politicians, and cultural producers (writers, artists). From a genetic point of view, admixed populations are useful for helping to trace interesting genetic variants, generally ones related to complex disorders; Latino populations (in the U.S.) have been suggested as a productive focus (Burchard et al. 2005). Recently, some geneticists have emphasized that genomic research is too focused on European-descendant people and would benefit from being more globally inclusive; genomic research could then benefit a wider population (Bustamante, De La Vega, and Burchard 2011). In this sense there is a dual scientific and moral argument for studying Latin American mestizo populations (among others; see also Fullwiley 2008).

Recentering the Mestizo and Mestizaje/Mestiçagem

The critique of genomics' Eurocentric focus took on a more postcolonially nationalist tone in the case of INMEGEN's project to map Mexican genetic diversity. In Brazil, Colombia, and Mexico, there has been great concern to protect the nation's genetic resources, including human (and especially indigenous) genomes, against biopiracy and foreign exploitation, but the idea here was a little different (see Benjamin 2009; Barragán 2011; Santos 2002). It was a matter of tailoring public health policy to the genetic characteristics of the Mexican population and not being overly bound by treatments generated by Eurocentric medical research and industry. A similar argument has been made for Brazil in relation to pharmacogenetic treatments (Suarez-Kurtz 2011). In Mexico, the notion of genomic sovereignty over a human genetic resource or territory seen as somehow distinctive became acceptable, despite the problematic implication that Mexicans, as a national population, had a genomic profile that was identifiably particular (Benjamin 2009; Schwartz-Marín and Silva-Zolezzi 2010). The mestizo thus circulated as an object of interest in genomics, health policy, nation building, and postcolonial pride, even as multiculturalist initiatives proceeded from the early 1990s.

In Brazil, debates about the nation, mixedness, and multiculturalism took a slightly different turn. There, the question of affirmative action, especially in relation to health policy and racial quotas for university admissions, became a privileged arena for the circulation of genomic knowledge about mixture (see introduction). Without seeking to prescribe social policy on the basis of genetic science, Sérgio Pena and others argued that policy should take note of the scientific evidence—which was that many Brazilians identifying as black or brown had plenty of European ancestry, that 87 percent of all Brazilians had more than 10 percent African ancestry, and that it was impossible to biologically demarcate racial groups (Pena and Bortolini 2004). The implication was that racial quotas in education would be arbitrary and unfair in practice. Critics of this argument replied that social policy is directed at social categories, not genetic ones, and that the biological incoherence of Brazilian color categories was simply irrelevant when it came to trying to correct the clearly racialized dimensions of social inequality in the country.

In relation to health, the genetic argument was that, using standard Brazilian color/race categories, policies could never target a population that had a common genetic profile that might relate to health. The categories overlapped too much in genetic terms and medicine therefore had to treat the individual in terms of his or her ancestry (Pena 2005). Likewise, policies directed at the black population, defined to include pretos (blacks) and pardos (browns), would likely be misdirected, as pardos in some samples were genetically closer to brancos than to pretos (R. V. Santos et al. 2009).

Underlying these debates was a sense that affirmative action based on race/

color was also seen as un-Brazilian, as a foreign import that did not suit the social realities of the country. In this context, the mestiço circulated as a figure testifying to the centrality of mixture for the Brazilian nation and—whatever the undeniable presence of racism and racially structured inequality—against the idea that enshrining racial/color difference in social policy was either effective practically or a good way forward politically. Again, this was in the context of ongoing multiculturalist reform, starting with constitutional changes in 1988 and proceeding through an official recognition of racism in the mid-1990s and a number of legal measures aimed at Afrodescendant and indigenous groups, of which racial university quotas and health policies were only the most hotly debated examples.

Colombia presents a rather different scenario. Multicultural reforms have been extensive by Latin American standards, and did include some affirmative action measures, which, however, have thus far fallen well short of Brazilian-style quotas and health policy interventions. Still, there has been a good deal of public debate on multiculturalism, which has included some of the same arguments as in Brazil—on the one hand, that enshrining racial and ethnic difference in policy runs counter to the nation's underlying character and history; and that, on the other hand, the new recognition of Afro-Colombian and indigenous minorities will increase equality and combat racism. However, genetics has not figured much in these debates—although it has figured in controversies about so-called gene hunting among indigenous peoples and the need to protect them as part of a multiculturalist regime (Barragán 2011). Colombia indicates how mestizaje and the mestizo can work as genomic objects, without being drawn so directly into heated political debates about particular multiculturalist policy issues.

For all three countries, in simple terms, it is striking how these genomic objects serve to recenter the mestizo majority in the nation and to reinscribe the process of mestizaje at the core of national identity. In some ways, genomic research powerfully reinscribes the categories that multiculturalism highlights—indigenous peoples, Afrodescendants—which of course were already well established, albeit under different names (e.g., indios, negros). But there is a very clear tendency to focus on the category that, in the words of the Colombian census of 2005, figures as "none of the above," that is, neither indigenous nor Afrodescendant. In other words, the focus is on what, in multiculturalism, remained the unmarked category—the mestizo. In addition, in Colombia and Brazil, there was a certain tendency to establish that the mestizo majority was also quite European in ancestry, within the diversity that genomic testing revealed. This came out partly in the projects that focused on specific regional populations seen as relatively white—the Antioqueños in Colombia, the Gaúchos in Brazil—but also in statements such as "Regardless of their skin colour, the overwhelming majority of Brazilians have a high degree of European ancestry" (Pena et al. 2009: 875) and "In all regions studied the European ancestry was predominant" (Pena et al. 2011: 5).[12]

Finally, as noted above, many of the studies produced data on indigenous mtDNA and European Y chromosome markers, explaining it with the gendered narrative of European men having sex with indigenous and African women. This story is foundational to the whole concept of mestizaje and reiterates both mixedness itself and the dominance of masculine whiteness in a number of ways. First, the use of a technical terminology of "asymmetrical mating" (or the nontechnical use of kinship terms such as "indigenous mothers") tends to gloss over the exploitation and violent abuse that undoubtedly formed part of these sexual encounters. Mestizaje appears implicitly as a peaceful and harmonious process. Second, the story selectively erases European women and indigenous and African men whose mtDNA and Y-DNA lineages did not survive unbroken into the samples of mestizos and brancos that often underlay this research. Other samples of African-derived populations in Brazil showed high levels of African ancestry in both mtDNA and Y-DNA, yet research tended to emphasize the "most consistent finding," which is "the introduction of European genes through males" (Bortolini et al. 1999: 560). The sexual agency and dominance of European men was placed in the foreground, alongside the sexual receptiveness of indigenous and African women.

Third, the indigenous and African women who appear as sexual partners —or in nontechnical terms as mothers—played this role once, long ago, in early colonial times. After this, their mitochondrial DNA was inherited and passed on by other women/mothers who were increasingly unlikely to have been either indigenous or African. The same goes for European fathers. After the early colonial period, most mixture would have taken place between mestizo men and women, rather than between indigenous or African women and European men. The focus on indigenous or African women/mothers and European men/fathers radically telescopes this history by capturing mainly early colonial processes and then generalizing them to narrate patterns of DNA. The trope of mixture between clearly distinctive categories is foregrounded, thus reiterating the founding narrative of mestizaje.

In short, genomic research in the three countries can be seen as a reassertion of Latin American mixture and thus also specificity. Multiculturalism in the region was undoubtedly driven in part by transnational processes of grassroots diasporic activism and international multilateral organizations (from the church and the UN to the Ford Foundation; Bourdieu and Wacquant 1999; Telles 2003; Wade 2010). Latin American states chose that route because of pressure from minorities and because this was now the definition of a modern liberal democracy—or a "neoliberal" one (Hale 2005). Genomic research highlighting mestizaje does not of course dismantle multiculturalism—it reinforces it in certain ways—but it does enunciate a Latin American particularity: the majority of the nation is mixed and thus has always already been multicultural in a characteristically Latin American and mixed fashion.

Citizenship and Genomic Research

What are the consequences of genomic research on mixedness and diversity in terms of citizenship and equality in a multicultural society? The major area for this discussion is the association of certain disorders with specific ancestries (Braun et al. 2007; Fausto-Sterling 2004; Fujimura, Duster, and Rajagopalan 2008; Montoya 2011). This is a significant way in which citizenship and inequality can be construed through a complex intersection of biology and culture. In the process of exploring possible connections between ancestry, populations, and disease, geneticists may contest the racialization of certain disorders: this happened in Brazil in relation to sickle cell anemia, a disorder whose transmission follows Mendelian patterns (Fry 2005b; Pena 2005). In the case of complex disorders, where the role of genetic causes is much less clear, things may turn out differently. Categorical labels of ethnicity and race may be applied, which generalize across populations such that, for example, in research into diabetes and Mexican American populations, "disparities in health are converted into biological differences," leading "the biomedical enterprise to pathologize ethnicity" (Montoya 2011: 185, 186). For Montoya, this is not a straightforward result of the unreflective use of race and ethnicity in biomedical research, as researchers generally do not believe that races are valid biological entities. Much less is it the result of racism by researchers, who aim to benefit minority populations. It is the result of a slippage between descriptive modes of talking about racial-ethnic populations, for example as Mexican or as having certain statistical patterns of biological variance, and attributive modes of talking, which link these populations to a genetic propensity to, say, diabetes (Montoya 2011: 184). The consequences of such pathologization are that the social determinants of poor health—and particularly of poor health for minority groups—are not given enough attention.

In addition, some scholars argue that associations of ancestry or ethnic identity with disease may entrain moral blame. For example, if you have high levels of Amerindian ancestry, you may be thought more liable to suffer from diabetes and be seen by others, within and outside the medical profession, to carry a greater moral responsibility to regulate your lifestyle in healthy ways. Perceived failure to meet that responsibility may label you, and others who share that kind of ancestry, a less worthy citizen (Rose and Novas 2005). On the other hand, the sharing of genetic traits associated with disease can also lead to citizen activism and the taking on of new responsibilities, including engagement with the state and biomedical enterprises (Fry 2005b; Heath, Rapp, and Taussig 2007).

Our research focused on a small number of case studies, which makes it unwise to generalize for whole countries, but judging on the basis of the genomic research we observed, these kinds of effects were only incipient. Despite the evident concern of the Mexican state with very high levels of diabetes and obesity in the country and despite research exploring links between these levels and genetic characteristics of the Mexican population, it is not evident that they are leading to differentials around citizenship within the country—although significant transnational aspects may be involved.[13] In Brazil, the situation is different insofar as some geneticists intervened in debates about affirmative action in higher education and health policy, which have clear consequences for citizenship. The tenor of the interventions was generally to challenge racialized affirmative action and, by implication, to support more universalist social policies—although, despite this, affirmative action programs remain in place. In this sense, some genomic science cast as unscientific both the black social movement and policies directed at blacks.

Overall, the effects on citizenship are more indirect and revolve around the construction of indigenous and Afrodescendant groups as located outside the core of the nation. The imagined community that emerges from genomic research is essentially a mestizo one, envisaged as one category, but simultaneously hierarchized by differential proportions of genetic parental ancestries. Indigenous and Afrodescendant groups, struggling for rights and equal treatment, then reappear as outsiders fighting to get in, rather than distinctive cultural sectors in a multicultural nation.

Comparisons

Comparison entails separation of one case from another, yet it requires a common frame within which the comparison makes sense. Foregrounding the frame brings into view the connections between the cases, in terms both of their common origins and of possible interactions between them. Foregrounding the cases themselves tends to hide commonality and interaction, isolating the cases as contrastive.

Regional Comparisons

In the study of race in the Americas, a classic comparison has been between the United States and Brazil (and by extension other areas of Latin America). It is a comparison widely made by academics and other intellectuals over the course of the twentieth century: a fundamental aspect has been the apparent contrast between a highly segregated society, founded on clear racial categories and very restricted processes of mixture across them, and a mixed society, less racially segregated and with blurred racial categories. The debate has gone back and forth on the nature and validity of this apparent contrast. But if the frame within which the comparison is taking place is foregrounded, then both cases appear as variants of common processes and, moreover, interactions between the cases are revealed, which in this case included precisely attempts to contrast the two countries by intellectuals who moved between them and had particular interests in making the distinction (Fry 2000; Marx 1998; Seigel 2009; Wade 2004; see also Hartigan 2013b).

Genomics occupies an interesting role here. It was evident from the start that, whatever the importance of national frames, genomic science was also a highly transnational enterprise. Scientists collaborated within Latin America and with North American, Asian, and European labs; they published in international as well as national journals; they used international data sets; and their data sets were used internationally. There was a broad transnational frame within which different labs operated to more or less common standards. In this frame, Latin American populations appear as a variant like any other population in the world. As one Colombian researcher put it, "From a genetic point of view, [mestizo] is a universal concept, it is just that it has been more consolidated in America."[14] From this perspective too, the attempt to link specific ancestries to specific diseases could be seen as part of a global enterprise: type 2 diabetes can be linked to genetic variants in South Asian populations, which also have high frequencies of the disease; the specificity of Amerindian ancestry recedes in this view.

On the other hand, as we have seen in this conclusion and chapter 7, some Latin American geneticists were clearly also interested in making the populations of the region distinctive, as Mexicans, as Brazilians, as Latin Americans, as South Americans, as mestizos. This might obey postcolonial motivations of securing genomic sovereignty or of combating Eurocentrism in sampling, data sets, and medical and pharmaceutical research (Burchard et al. 2005; Bustamante, De La Vega, and Burchard 2011; Suarez-Kurtz 2011). It also fed into depictions of individual countries as essentially mestizo and thus implicitly as different from the United States.

In this sense, genomics was playing out long-standing comparisons of race relations in North and South America. This was not a stark contrast between the overriding importance of race in the North versus its insignificance in the South; it was simply a strong assertion of the overwhelming fact of genetic mixture as distinctively Latin American. As the Brazilian geneticist cited in chapter 7 put it: "Miscegenation is good, because the alternative is segregation."[15]

National Comparisons

In our project, the work of comparison also sought to disaggregate the idea of a Latin American mestizaje. We started with a hypothesis of national differences, despite our awareness of the transnational dimensions of genomic science. At first sight, there seemed to be some interesting differences between the three countries. The apparent genomic nationalism of Mexico, with its state-funded institute and its explicit concern with genomic sovereignty and national health policy fit with the strong institutionalization of the mestizo as the core national identity and the long-standing ambivalence of relationships with its big northern neighbor. The irreducibility of the indígena category in genomic research, as the counterpoint to the mestizo, fitted with the political ideologies and practices of indigenismo that have been particularly powerful in Mexico. The emphasis in Brazil on everyone being genetically mixed, on the nonexistence of races, and on the presence of African ancestry in most of the population squared well with the celebration since the 1930s of a characteristically Brazilian triethnic mix, which supposedly avoided the racial traumas of the United States and included cultural elements labeled as having black Brazilian origins (such as samba) as national icons. In Colombia, there seemed to be a greater emphasis on internal genetic diversity and regional distinctiveness, which meshed with long-standing representations of Colombia as a "country of regions," whether this was seen as problematic fragmentation or delightful diversity.

In some ways, these initial impressions were borne out, as the chapters in this book indicate, but the methodological nationalism that separated the three countries as distinctive cases hid other processes at work. The first, already explored above, was the common, transnational frame for genomic science in all three countries, which also drew the scientists together into international networks between Latin American countries. There was also a common frame in the sense of highlighting Latin American populations, in their variety, as essentially mixed and therefore distinctive in the Americas (and perhaps globally). There was the consistent emphasis on the gendered pattern of indigenous (and African) mtDNA and European Y chromosome DNA. All three countries were concerned with protecting genomic sovereignty or genetic patrimony, even if this was more institutionalized in Mexico, via INMEGEN. In all three countries, the category indígena or Amerindian carried connotations of isolation and separation, linked to ideologies of indigenismo, even if the situations of indigenous peoples and the historical importance of indigenismo were rather different in each place. In short, the countries were linked by sharing common histories in a global frame.

The second, perhaps more interesting dimension, hidden by national comparisons, was the variety within each country. This was most evident in Brazil, where genomic research was well developed and spread across many different labs. Geneticists such as Sidney Santos, working in Belém, on the Amazon, and Maria Cátira Bortolini, based in the southern city of Porto Alegre, focused on the specificity and internal diversity of populations located in the regions where they worked, which in the case of Bortolini was not just a regional category but a named group, the Gaúchos, whose members considered themselves ethnically distinct (Marrero et al. 2005, 2007; Rodrigues, Palha, and Santos 2007; Santos, Rodrigues, et al. 1999). Different labs presented multiple versions of Brazil.

Diversity of genomic research was less evident in Mexico, where INMEGEN tended to dominate the landscape, having secured a position as an obligatory passage point for genomic research in the country. Still, a different dimension of internal variety was evident in the way INMEGEN itself changed over time. As related in chapter 3, under its first director, Gerardo Jiménez Sánchez, INMEGEN focused on creating "a national genomics." When Xavier Soberón took over as director in 2009, the emphasis changed: there was less emphasis on genomic sovereignty and greater interest in international collaborations, using the data sets generated by the Mexican Genome Diversity Project (which were already being accessed primarily by non-Mexican researchers); there was also more focus on sequencing whole indigenous genomes in the search for rare variants. The reasons for these shifts are complex, but López Beltrán, García Deister, and Rios Sandoval (chapter 3) conclude that "the centralized and distinctively nationalist character" that Jiménez Sánchez gave to INMEGEN had no place "within the reticular and large consortium-prone structure of genomics research at large." In any case, it is clear that it was not simple to talk about a Mexican pattern of genomics, nor a single way in which Mexican genomics represented the nation.

In Colombia, there was also significant diversity. In Medellín, there was a good deal of interest in the local Antioqueño population, seen as a "population isolate," with a distinctive genealogical and genetic history (see chapter 7). This lab was well placed in international networks of genomic science. In Bogotá, our project worked with a lab led by more junior researchers, less well connected internationally, and with a consciously self-reflective approach to the contingency of genomic knowledge and population categories (see chapter 5). Many researchers agreed in principle on the idea of Colombia as a regionally diverse country, but more or less emphasis could be laid on the specificity of certain regional populations.

What this experience of comparison tells us is that, although nation or Latin America may have been a frame of reference for some of the genomic science, these entities were constantly being traversed by commonalities across nations or larger regions and by diversity within them, with both aspects being orchestrated in part by the transnational character of genomic science. There is no scale of resolution at which an irreducible unit of comparative analysis can be taken for granted; we must always be alert to the interconnections of the networks in play.

Final Reflections

What interventions do we hope to make with this project? First, our research tries to elicit the way genomics refigures race in quite specific ways. We seek to go beyond the idea that race simply reappears in genomic work—after all, race arguably never completely disappeared in the first place (Reardon 2005)— and show the effects of specifically genomic technologies. For example, the individualization of ancestry measurements, which produces an unlimited finity of differences and belongings, breaks down simple racial categories, while remaining structured by key ancestral reference points that are race-like in their evocations. Or the way that mtDNA and Y chromosome analysis both multiplies concepts of ancestry and genders them. It is important as well to grasp how genomic research both stabilizes and simultaneously destabilizes race and race-like ideas of biogeographical ancestry.

Second, we want to highlight the way ideas about (ad)mixture circulate in genomic research and debates about the character of the nation: in the context of multiculturalist political reforms and social movements, it is important to recognize how a significant body of genomic science is foregrounding mixture and the mestizo and thus recentering imaginaries of the nation around a genomically naturalized idea of mestizaje or mestiçagem.

Third, it is important to reflect on Latin America. This regional construct figures as a place to bring into high relief how genomic technologies provide a new molecular and bioinformatic idiom for racialized thinking. It is also a place that appears as home to a notion of mixture, which can act as a source of global capital both in genomic research (Latina/o Americans as scientifically interesting populations) and political multiculturalism (Latin America as having been always already multicultural in its mixed diversity and tolerance of mixture). Last, Latin America loses its coherence as we pay attention to transnational circulations of knowledge and internal multiplicity of practice. Our focus on specific genetic labs and projects, albeit a small selection of all those that exist, shows how new versions of race are being assembled globally through local practices.

Notes

1   The literature on this topic is growing fast. See, among others, Abu El-Haj (2007, 2012), Bliss (2009a, 2011), Condit et al. (2004), Condit, Parrott, and Harris (2002), Fry (2005b), Fullwiley (2007b, 2008), Gaspar Neto and Santos (2011), Gibbon, Santos, and Sans (2011), Koenig, Lee, and Richardson (2008), M'charek (2005a), Maio and Monteiro (2010), Montoya (2011), Reardon (2005), Santos and Maio (2005), Schramm, Skinner, and Rottenburg (2011), Skinner (2006, 2007), Wade (2002b), Whitmarsh and Jones (2010).

2   Not all labs produced percentage admixture estimates. Some labs in Colombia, while they generated statistical tables with allelic frequencies, resisted the idea of percentage ancestral contributions. As one researcher argued, unless they were able to define what being 100 percent African (or European or Amerindian) meant, it was not realistic to obtain percentages: "No one is pure. There is no 100 percent. . . . You are always analyzing a portion of the genome; therefore, a percentage is false" (interview data from Olarte Sierra and Díaz del Castillo).

3   For reference populations, Pena used the HGDP-CEPH Human Genome Diversity Cell Line Panel, which has samples from various African, European, and Amerindian populations (Bastos-Rodrigues, Pimenta, and Pena 2006). Silva-Zolezzi et al. (2009) used HapMap samples from Nigeria and Utah, plus their own samples of Zapotec individuals. Some of the graphic representations could not be reproduced in this book; e.g., the graphics in Silva-Zolezzi et al. (2009).

4   See, for example, "El 85,5 por ciento de las madres colombianas tiene origen indígena," El Tiempo, 12 October 2006, http://www.eltiempo.com/archivo/documento/CMS-3283433; Liliana Alcántara, "Listo, el mapa genómico de los mexicanos," El Universal, 9 March 2007, http://www.eluniversal.com.mx/nacion/149089.html.

5   http://www.facebook.com/CandelaMx.
6   On precision, science, and nation building more generally, see Wise (1997c).
7   See, for example, the case of the 2007 Afro-Brazilian Roots project, by the Brazilian Service of the BBC, which published the ancestry test results of nine celebrities (Salek 2007). Sérgio Pena also analyzed the DNA of two of Gilberto Freyre's grandchildren, concluding that he was the descendant of Sephardic Jews and sparking interest from the press (Santos and Maio 2004: 374). In the United States, Henry Louis Gates's 2012 TV series, *Finding Your Roots*, also used genetic ancestry testing to uncover the hidden roots of celebrities.
8   This phrase was coined by Vivette García Deister.
9   That categorizations are contingent on methods, even within regimes of standardization, is not new: "We can identify 'clusters' of populations [in the world] . . . [but] minor changes in the genes or methods used shift some populations from one cluster to another" (Cavalli-Sforza, Menozzi, and Piazza 1994: 19). This sense of contingency is intensified with genomic technologies.
10 See the map produced by William Usaquén at Universidad Nacional de Columbia, shown in Barragán Duarte (2007).
11 See the reports on this kind of research by Yunis: Rick Kearns, "Indigenous founding mothers of the Americas," CAC Review, 18 April 2007, http://cacreview .blogspot.com/2007/04/indigenous-founding-mothers-of-americas.html; "El 85,5 por ciento de las madres colombianas tiene origen indígena," El Tiempo, 12 October 2006, http://www.eltiempo.com/archivo/documento/CMS-3283433.
12 For the 2009 article, the data presented are mostly for self-classified whites; the black sample has 37 percent European ancestry (blacks—pretos in the census— are about 6 percent of the national population). For the 2011 article, the predominance of European ancestry is in terms of ancestral contributions averaged for samples of individuals (combined white, brown, and black) in four major regions.
13 If Mexican immigrants and Mexican Americans are blamed in some way for their own ill health, as Montoya (2011) contends, this could have implications for immigration policies as well as treatment within the United States.
14 Interviewed by Olarte Sierra and Díaz del Castillo.
15 Interviewed by Kent.

appendix

Methods and Context

The texts below give a very brief outline of (a) the scope of research in genetics and especially human population genetics in each country; and (b) the range of work undertaken by the project team in each country. The case studies presented in chapters 4–6 represent only particular aspects of that range of work.

Genetics Research in Brazil
Michael Kent and Ricardo Ventura Santos

In Brazil, human population genetics is a well-established academic field that has been growing steadily since the 1950s (Salzano 2011). At present, there are twenty-six state-recognized postgraduate programs in genetics, twenty-one of which include a PhD. The vast majority of these programs are located in public universities. The yearly conference of the Sociedade Brasileira de Genética normally attracts between 3,000 and 4,000 delegates. Research is mostly conducted with public funds, in particular via CAPES and CNPQ, the two main federal research councils, as well as the research councils of individual states in Brazil. There are high levels of internationalization, with much collaboration with research groups in other countries of Latin America, the United States, and Europe, as well as many publications in international journals.

Due to the size of the genetic field in Brazil, we chose to concentrate our attention on human population genetics, and in particular on research that explores the ancestry and formation of Brazilian populations. This area of genetics in itself constitutes a relatively large and autonomous field. Research on the genetic constitution of the Brazilian population has gained high levels of public visibility in the past decade (see chapters 1 and 4).

We identified three laboratories of particular importance. These laboratories are located, respectively, at the Universidade Federal de Minas Gerais (UFMG) in Belo Horizonte, coordinated by Sérgio Pena; at the Universidade Federal do Rio Grande do Sul (UFRGS) in Porto Alegre, the institution where Francisco Salzano and Maria Cátira Bortolini work; and at the Universidade Federal do Pará (UFPA) in Belém, where we have focused on the work of Sidney Santos and associates. In these laboratories, geneticists have conducted research both on indigenous populations and early migration routes into the Americas, and on the constitution of nonindigenous populations of Brazil since the start of colonization in 1500. These two domains in practice operate as relatively separate fields. Research with indigenous samples mostly focuses on migration routes and long-term microevolutionary processes. In studies of mixed populations, indigenous populations appear mainly as parental populations. As our key interests were in race, mixture, and nation, we focused mainly on the latter studies. However, where relevant, we have also paid attention to studies conducted on indigenous populations, as well as research in medical and forensic genetics.

Each of the three laboratories is situated in a different macroregion of Brazil, respectively the southeast, the south, and the north (which comprises most of the Amazon region). These regions are perceived as distinct in sociocultural and phenotypical terms, with relatively more European elements in the south and indigenous ones in the north, with the southeast often considered as representative of Brazil as a whole. The three laboratories adopt slightly different approaches to the study of ancestry in Brazil, giving rise to different images of the genetic constitution of the nation (see the conclusion). Located in the southeastern region, Sergio Pena's work focuses mainly on the national level and on the genetic aspects that unite the Brazilian population, while researchers at UFPA in the north and UFRGS in the south have paid particular attention to regional differentiation within Brazil, although researchers from these two institutions also participate in projects aimed at characterizing the genetic makeup of the Brazilian population as a whole. The choice to divide research efforts over three different laboratories allowed us to explore the variety of genetic studies of Brazilian populations.

All three laboratories have strong ties with foreign research groups in the United States, the United Kingdom, Argentina, Colombia, Mexico, Germany, and Portugal. However, while publications relating to indigenous populations tend to be international collaborations, articles that focus specifically on the ancestry and formation of other contemporary Brazilian populations tend to be authored mostly by Brazilian researchers. In all three laboratories, researchers tend to publish their articles on the ancestry of contemporary Brazilians in leading journals in the fields of biological anthropology and human genetics, such as *American Journal of Physical Anthropology*, *American Journal of Human Biology*, and *American Journal of Human Genetics*, among others.[1]

The funding of research in these laboratories generally follows the pattern noted above. The laboratory at UFPA gains additional funds from a contract with the public prosecutor's office to carry out paternity tests. In addition to his work at the laboratory of UFMG, Sergio Pena also owns a private company—Laboratório Gene—that has been providing a variety of genetic tests on a commercial basis since the 1980s.

For the present project, contacts with researchers at the three laboratories were established by Ricardo Ventura Santos, drawing on earlier exchanges and collaborations with these researchers. Access for ethnographic fieldwork by

Michael Kent was negotiated at the laboratories at UFRGS and UFPA. Santos and Kent conducted an extensive interview with Sergio Pena, and a great deal of data on Pena's work is available through his own books and publications and appearances in written and audiovisual popular media. The participation of a Brazilian research assistant for the current project, Verlan Valle Gaspar Neto, was focused predominantly on systemizing and analyzing this material (see Gaspar Neto and Santos 2011; Gaspar Neto, Santos, and Kent 2012).

The laboratory ethnographies Kent conducted at UFRGS and UFPA involved a variety of methodological approaches, including accompanying the everyday routine of researchers; conducting informal conversations and recorded interviews; participating in classes and debates; and analyzing the laboratories' scientific production such as journal articles, theses, and research reports. Such engagements involved both established academic staff members and postgraduate researchers. These ethnographies were roughly divided into two phases. During the first phase, fieldwork focused broadly on the mapping and analysis of the diverse studies conducted at these laboratories that were relevant to the central theme of the present project—the articulation between genetic research on human populations and ideas about ethnicity, race, mixture, and nation—as well as the geneticists' perspectives on these issues. This phase of research was conducted at the UFRGS in the months of June and July 2010, and at the UFPA in August of the same year.

During the second phase of fieldwork—in October and November 2010—Kent focused on a specific case study in each laboratory. At the UFRGS the chosen case consisted of a research project on the regional genetic profile of the Gaúchos, the inhabitants of Rio Grande do Sul's rural Pampa region, with a particular focus on the Gaúchos' connection with the indigenous Charrua. This research is presented in chapter 4. Finally, the case study conducted at the UFPA focused on the articulation between the expansion of the laboratory's resources and the shift of its focus of research from the Amazonian population to the Brazilian population as a whole. As it has only been possible to include one case study in this book, this specific case will be published elsewhere. However, some of the evidence it generated is discussed in the comparative analysis of laboratory categories and practices in chapter 7.

Genetics Research in Colombia
María Fernanda Olarte Sierra and Adriana Díaz del Castillo H.

In Colombia, genetics research has three axes: population genetics, medical genetics, and forensic genetics. Population genetics research is mainly conducted by academic laboratories of different universities across the country. Research is not necessarily linked to state agencies, but there is some collaboration. Medical genetics research is also conducted in public institutions such as the National

Health Institute, while forensic genetics is intertwined with legal organizations such as the Fiscalía (public prosecutor's office) or, for cases of paternity testing, the Comisarías de Familia (family welfare agencies). Funding for academic research comes from Colciencias (the state science funding agency), the universities, and Banco de la República (the state bank). Public institutions such as the National Health Institute may also receive state funding. All human genetics research is based on the understanding that the Colombian population is triethnic and that levels of admixture vary from region to region (Builes et al. 2004; Rojas et al. 2010; Yunis et al. 2005; Yunis Turbay 2009).

Building on National Identity: Population Genetics

Research on population genetics is seen as contributing to the understanding of continental and national population histories and processes of admixture (Bedoya et al. 2006; Mesa et al. 2000; Rey et al. 2003; Ruiz-Linares et al. 1999). Population geneticists have produced information regarding ancestry, and great emphasis is placed on the ancestral composition of individuals, based on the study of mtDNA, Y-ChDNA, and autosomal DNA. Labs generally produce their own data but may also use data published by fellow Colombian geneticists such as Paredes et al. (2003), as well as data produced by other Latin American geneticists. After the 2008 National Congress on Genetics, geneticist William Usaquén tried to initiate a national population genetics database, to allow a clearer understanding of variations in genetic makeup within the country. Geneticists welcomed this effort, but so far no concrete measures have been undertaken.

We found that all population genetic studies shared the idea of a regionalized national geography, which we argue has racialized connotations. However, this basic regionalization has varied ways of being enacted, depending on each lab and sometimes on each geneticist (see Olarte Sierra and Díaz del Castillo 2013). These studies are considered by the geneticists as valuable means for enhancing national identity and for reducing racism. It is expected that scientific evidence that all are mixed will undermine racist attitudes.

Medical Genetics: A Healthier Future

The promise of a healthier future is a starting point for many geneticists in Colombia. Medical genetics receives an important share of genetic research funding. Research in medical genetics ranges from the study of how admixture plays a role in the expression of diseases such as diabetes, hypertension, metabolic syndrome, and Alzheimer's, to mention just a few, to genetic epidemiology, which focuses on finding the genetic basis of different types of cancer, depression, multiple sclerosis, and genotoxicity. Some laboratories perform cytogenetic analysis and do genetic diagnosis of a number of biochemical disorders. In this case, as well, the role of admixture is considered paramount.

Forensics: The Need for Colombia-Specific Genetic Markers

Forensic genetics, an important activity for geneticists in Colombia, is the main source of local genetic profiles. Forensic genetics has contributed to the development of genetics in general, through the use of new technologies and through the constant generation of data.

Forensic genetics in Colombia began with paternity testing and expanded with the identification of the bodies of victims who had disappeared during long-lasting armed conflicts. The Colombian Institute of Family Welfare (Instituto Colombiano de Bienestar Familiar, ICBF) has been in charge of adjudicating paternity. In the early 1960s, this was based on anthropometric measures. From 1972 to 1998, blood group identification was used and, from 1995, DNA tests. Nevertheless, there was no reference population against which results could be compared. To genetically identify a person, the individual genetic profile (allele frequencies) had to be compared to the allele frequencies of a given reference population in order to establish the probability that such a profile might belong to another individual in the population (M'charek 2005a).

In 2003, Manuel Paredes and collaborators from the National Institute of Legal Medicine and Forensic Sciences in Colombia and the Institute of Legal Medicine at the Universidad de Santiago de Compostela in Spain published an article in *Forensics Science International*, presenting a table of allele frequencies for the Colombian population. Data were gathered from 1,429 samples across the country and were organized into four regions. The ICBF decided that the Paredes table would be used as the official one for solving paternity cases. The Institute of Legal Medicine and the public prosecutor's office then adopted the table for all other forensic cases, including those related to the 2005 Peace and Justice Law (Usaquén, personal communication, 2010), which aimed to compensate victims of the armed conflict and to encourage the demobilization of illegal armed forces and which created a section of the public prosecutor's office dedicated to the exhumation of victims' bodies.[2]

Our Research in Colombia

We researched four human genetics laboratories.[3] In laboratory A we conducted an eight-month ethnography; laboratory B constituted a follow-up case; and laboratories C and D were studied through in-depth interviews. The different methodological approaches responded mainly to our access to the laboratories, the geneticists working there, and the type of research they conducted. The funding of these laboratories comes from the university they belong to, Colciencias, and, occasionally, the state bank.

Laboratory A is the Genetics Institute of the Universidad Nacional de Colombia in Bogotá; specifically its research group, Genetics of Population and Identification. An aspect of the work of this institute is presented as a case study in chapter 5.

Laboratory B belongs to the science faculty of a public university. It has a strong interest in characterizing and quantifying admixture in the population of the region where the lab is located.[4] Researchers are interested in genetic epidemiology, paying special attention to the role of admixture in the expression of diseases and in the population history of the country. This lab has over fifteen years of experience, reflected in many publications and projects. It has more than fifteen students (undergraduate, master's, doctoral) and about fifteen researchers specializing in different areas. Some of its members travel abroad for training, as this lab has partnerships with other institutions. Foreign institutions fund some projects.

Laboratory C is part of a private university and is smaller than the other three. However, it is involved in many diverse projects. Researchers are interested in the genetic epidemiology of cancer and were previously interested in population genetics. The practice of the geneticists we interviewed is shaped by a vision of science as neutral, objective, and not necessarily influenced by other disciplines.

Laboratory D is part of a private university. It is the only lab we studied that functions as part of a school of medicine and its researchers are mainly physicians. Several groups work on medical genetics, with a small group conducting human population genetic studies, collaborating with researchers in other Colombian universities. At the time of writing, they are using ancient DNA to explore historical and archaeological theories about the peopling of the continent and the country. Also, they work on genetic epidemiology of cancer, comparative genetics (with other species), and the history of genetics and science.

The Landscape of Human Genetics in Mexico
Vivette García Deister and Carlos López Beltrán

The landscape of human genetics in Mexico is rich in terms of discipline, objectives, methodologies, objects of study, funding bodies, and institutions. As with other Latin American countries, this variety can be organized around three main axes of investigation: medical genetics, forensic genetics, and molecular anthropology. All of these approximations involve the practice of population genetics, which encompasses theoretical, laboratory, and field work.

Medical genetics was the first branch of human population genetics to be developed in Mexico. During the second half of the twentieth century it relied on serological and immunological techniques, but the discipline's methodological repertoire has been constantly updated. Most of the current research on medical genetics, that is, the identification of genes implied in the susceptibility and development of common diseases in the Mexican population, is centralized and carried out in National Institutes of Health or government hospitals. Two of the main research projects on metabolic syndrome (obesity and diabetes) that involve genomic sequencing take place in the Salvador Zubirán National Institute of Medical Sciences and Nutrition (Instituto Nacional de Ciencias Médicas y Nu-trición Salvador Zubirán) and in the Twenty-First-Century Medical Center (Cen-tro Médico Siglo XXI), in Mexico City. Two sample populations are used in these investigations: Mexican mestizos and *indígenas* (indigenous people). The funding for medical genetics projects comes from the National Council of Science and Technology (Consejo Nacional de Ciencia y Tecnología) and state-funded public universities. Private foundations also play a role. The Slim Initiative for Genomic Medicine serves as a liaison between Mexico's National Institute for Genomic Medicine (INMEGEN) and MIT's Broad Institute, and provides financial support for research on diabetes and cancer. Since 2008, the Coca-Cola Foundation has awarded 8.1 million pesos (about 650,000 USD) to the School of Chemistry of the National Autonomous University of Mexico (Universidad Nacional Autónoma de México, UNAM). These funds have been instrumental in the creation of a library of indigenous gene variants (the *genoteca indígena*). The object of this project is to identify genetic variants of indigenous ancestry that might be associated with the onset of diabetes in the Mexican mestizo population.

Forensic genetics is still an underdeveloped branch of human genetics in Mexico. Profiling by DNA is used in paternity testing and identification of vic-tims of crime or natural disasters. According to official figures, nineteen of the thirty-two states that make up the Mexican Republic have an onsite genetics laboratory, where samples are collected but not always processed. The Attorney General's Office (Procuraduría General de Justicia, PGR) in Mexico City attracts most of the cases associated with organized crime and processes the samples obtained from other states. Although the FBI's CODIS was implemented in 2011, the PGR is still in the process of creating a national database of genetic profiles. By 2013, less than 500 bodies had been identified using techniques of forensic genetics, whereas the number of unidentified victims and missing persons could be counted in tens of thousands.

Molecular anthropology (or anthropological genetics) is mostly concerned with determining evolutionary links between ancient and modern human popu-lations. Of particular interest in Mexico is the identification of migration patterns in pre-Hispanic populations and dynamics of admixture during the colonial pe-riod. Its main objects of study are ancient DNA recovered from archaeological and historical skeletal material, and contemporary DNA obtained from donors classified either as mestizo or indígena. Handling ancient DNA is methodologi-cally very tricky, as samples are often contaminated or in poor shape (degraded), requiring highly specialized laboratories. One of Mexico's most advanced ancient DNA laboratories is located in the National Laboratory of Biodiversity Genomics (Laboratorio Nacional de Genómica para la Biodiversidad) in the state of Guana-juato. Analysis of contemporary DNA samples is more widespread. Some of the institutions where this kind of research takes place are the National School of Anthropology and History (Escuela Nacional de Antropología e Historia), Center for Advanced Research and Study (Centro de Investigación y Estudios Avanzados), and several research centers in UNAM. Molecular anthropologists often cooperate with medical geneticists not only in Mexico but also abroad.

The increasing participation of bioinformatic tools in human genetics research prompted the creation of UNAM's Center of Genomic Science (Centro de Ciencias Genómicas, CCG) in 2004. Here future researchers in the field of genomic sciences are educated and trained. Recruitment of genomicists and bioinformaticians by research institutions and the government intensified from about 2007, and CCG graduates have begun to occupy these jobs.

Given this broad range of human genetics, why did we choose to focus on INMEGEN? The scale of ambition and notoriety that INMEGEN's Mexican Genome Diversity Project (MGDP) achieved between 2004 and 2009 sharply contrasted with the landscape of genomics research described here. INMEGEN portrayed a sense of novelty and profound difference with respect to the efforts accumulated over several decades of genetic research of Mexican populations, and aimed at taking over a space that at least some medical geneticists and molecular anthropologists were willing to contest. For this very reason, and because INMEGEN researchers were kind enough to welcome us into their laboratories, we focused on this institution and project. Also, investigating the birth and development of the MGDP provided a unique opportunity to understand the ways in which categories of race, ethnicity, and nation are used by Mexican scientists in the medical genomics scene.

Notes

1   These researchers have also published articles, mostly on indigenous populations and medical genetics, in leading journals in general sciences, such as *Nature*, *Proceedings of the National Academy of Sciences*, and *Science*.

2   Article 48, Law 975, 2005.

3   We name only the lab with which we worked most closely. We have not named the other three labs, as one of them requested to remain anonymous.

4   Admixture here refers to a person's percentage of African, Amerindian, and European ancestry.

references

Abu El-Haj, Nadia. 2007. The genetic reinscription of race. *Annual Review of Anthropology* 36(1): 283–300.

Abu El-Haj, Nadia. 2012. *The genealogical science: The search for Jewish origins and the politics of epistemology*. Chicago: University of Chicago Press.

Acuña-Alonzo, Víctor, Teresa Flores-Dorantes, Janine K. Kruit, et al. 2010. A functional ABCA1 gene variant is associated with low HDL-cholesterol levels and shows evidence of positive selection in Native Americans. *Human Molecular Genetics* 19(14): 2877–2885.

Agassiz, Louis, and Elizabeth Cabot Cary Agassiz. 1879. *A journey in Brazil*. Boston: Houghton, Osgood.

Agostoni, Claudia. 2011. Strategies, actors, promises and fears in the smallpox vaccinations campaigns in Mexico: From the Porfiriato to the Post-revolution (1880–1940). *Ciência e Saúde Coletiva* 16(2): 459–470.

Aguilar Rivera, José Antonio. 2001. Ensoñaciones de unidad nacional: La crisis en la identidad nacional en México y Estados Unido. *Política y Gobierno* 8(1): 195–222.

Aguirre Beltrán, Gonzalo. 1946. *La población negra de México, 1519–1810: Estudio etno-histórico*. Mexico City: Ediciones Fuente Cultural.

Aguirre Beltrán, Gonzalo, and Ricardo Pozas Arciniega. 1981. *La política indigenista en México*. Mexico City: Instituto Nacional Indigenista.

Alberro, Solange. 2006. *Del gachupín al criollo o de cómo los españoles de México dejaron de serlo*. Mexico City: El Colegio de México.

Alves-Silva, Juliana, Magda da Silva Santos, Pedro E. M. Guimarães, et al. 2000. The ancestry of Brazilian mtDNA lineages. *American Journal of Human Genetics* 67: 444–461.

Anderson, Benedict. 1983. *Imagined communities: Reflections on the origin and spread of nationalism*. London: Verso.

Andrews, George Reid. 1991. *Blacks and whites in São Paulo, Brazil, 1888–1988*. Madison: University of Wisconsin Press.

Anthias, Floya, and Nira Yuval-Davis. 1992. *Racialized boundaries: Race, nation, gender, colour and class and the anti-racist struggle*. London: Routledge.

Anzaldúa, Gloria. 1987. *Borderlands/la frontera: The new mestiza*. San Francisco: Aunt Lute.

Appelbaum, Nancy P., Anne S. Macpherson, and Karin A. Rosemblatt. 2003a. Introduction: racial nations. In *Race and nation in modern Latin America*, edited by Nancy P. Appelbaum, Anne S. Macpherson and Karin Alejandra Rosemblatt, 1–31. Chapel Hill: University of North Carolina Press.

Appelbaum, Nancy P., Anne S. Macpherson, and Karin A. Rosemblatt, eds. 2003b. *Race and nation in modern Latin America*. Chapel Hill: University of North Carolina Press.

Aréchiga Córdoba, Ernesto. 2009. Educación, propaganda o "dictadura sanitaria": Estrategias discursivas de higiene y salubridad públicas en el México posrevolucionario, 1917–1945. *Estudios de Historia Moderna y Contemporánea de México* 33(033): 57–88.

Arellano, Jorge, Martha Vallejo, Javier Jimenez, et al. 1984. HLA-B27 and ankylosing spondylitis in the Mexican Mestizo population. *Tissue Antigens* 23(2): 112–116.

Arias, William Hernán, Winston Rojas, Sonia Moreno, et al. 2006. Estructura y composición genética de una población antioqueña con alta prevalencia de Parkinson juvenil. *Salud uis, Revista de la Facultad de Salud, Universidad Industrial de Santander* 38: 56–57.

Azevedo, Eliane S. 1980. Subgroup studies of black admixture within a mixed population of Bahia, Brazil. *Annals of Human Genetics* 44(1): 55–60.

Balibar, Etienne. 1991. Is there a "neo-racism"? In *Race, nation and class: Ambiguous identities*, edited by Etienne Balibar and Immanuel Wallerstein, 17–28. London: Verso.

Banton, Michael. 1987. *Racial theories*. Cambridge: Cambridge University Press.

Barahona, Ana. 2009. *Historia de la genética humana en México, 1870–1970*. Mexico City: Universidad Nacional Autónoma de México.

Barkan, Elazar. 1992. *The retreat of scientific racism: Changing concepts of race in Britain and the United States between the world wars*. Cambridge: Cambridge University Press.

Barnes, Barry, and John Dupré. 2008. *Genomes and what to make of them*. Chicago: University of Chicago Press.

Barragán, Carlos Andrés. 2011. Molecular vignettes of the Colombian nation: The place(s) of race and ethnicity in networks of biocapital. In *Racial identities, genetic ancestry, and health in South America: Argentina, Brazil, Colombia, and Uruguay*, edited by Sahra Gibbon, Ricardo Ventura Santos, and Mónica Sans, 41–68. New York: Palgrave Macmillan.

Barragán Duarte, José Luis. 2007. Mapa genético de los colombianos. un *Periódico* 105. Accessed 8 January 2012. http://historico.unperiodico.unal.edu.co/ediciones/105/15.html.

Bartra, Roger. 2001. *Cultura y melancolía: Las enfermedades del alma en la España del Siglo de Oro*. Barcelona: Anagrama.

Bartra, Roger. 2005. *Anatomía del mexicano*. Mexico City: Debolsillo.

Basave Benítez, Agustín Francisco. 1992. *México mestizo: Análisis del nacionalismo mexicano en torno a la mestizofilia de Andrés Molina Enríquez*. Mexico City: Fondo de Cultura Económica.

Bastide, Roger, and Florestan Fernandes. 1955. *Relações raciais entre negroes e brancos em São Paulo*. São Paulo: Editora Anhembí.

Bastos-Rodrigues, Luciana, Juliana R. Pimenta, and Sergio D. J. Pena. 2006. The genetic structure of human populations studied through short insertion-deletion polymorphisms. *Annals of Human Genetics* 70(5): 658–665.

Becker, Itala Irene Basile. 2002. *Os índios Charrua e Minuano na antiga banda oriental do Uruguai, Série acadêmica*. São Leopoldo: Editora Unisinos.

Bedoya, Gabriel, Patricia Montoya, Jenny García, et al. 2006. Admixture dynamics in Hispanics: A shift in the nuclear genetic ancestry of a South American population isolate. *Proceedings of the National Academy of Sciences* 103(19): 7234–7239.

Bellagio Initiative. 2005. Genome-based research and population health: Report of an expert workshop held at the Rockefeller Foundation Study and Conference Centre in Bellagio, Italy, 14–20 April 2005. Public Health Genetics, University of Washington School of Public Health, Centers for Disease Control and Prevention.

Benjamin, Ruha. 2009. A lab of their own: Genomic sovereignty as postcolonial science policy. *Policy and Society* 28(4): 341–355.

Benzaquen de Araújo, Ricardo. 1994. *Guerra e paz: Casa Grande e Senzala e a obra de Gilberto Freyre nos anos 30*. Rio de Janeiro: Editora 34.

Bernal, Jaime. 1989. Editorial. *Boletín Expedición Humana* 1992, no. 1: 1.

Bernal, Jaime. 1991a. Expedición Humana 1992—a la zaga de la América Oculta: Estudios antropo-genéticos en poblaciones aisladas colombianas. Unpublished manuscript (research proposal submitted to Colencias).

Bernal, Jaime. 1991b. Que me ha dado la Expedición Humana. *Boletín Expedición Humana* 1992, no. 9: 2.

Bernal, Jaime. 1993. Hay mucho más en nuestro Aleph: Acto de Clausura de la Gran Expedición Humana. *América Negra* 6: 153–156.

Bernal, Jaime. 1996. Carta a Luis Guillermo Vasco. *Kabuya: Crítica Antropológica* 2: 6–7.

Bernal, Jaime. 2000. Presentación. In *Variación biológica y cultural en Colombia*. Vol. 1, *Geografía humana de Colombia*, edited by Adriana Ordóñez Vásquez, 9–21. Bogotá: Instituto Colombiano de Cultura Hispánica.

Bernal, Jaime. 2013. Los estudios genéticos de la Expedición Humana: ¿Cuál es su importancia? Expedición Humana. Accessed 5 April. www.javeriana.edu.co/Humana/cifras.html.

Bernal, Jaime, and Marta Lucía Tamayo. 1994. *Instituto de Genética Humana (1980–1994)*. Bogotá: Pontificia Universidad Javeriana.

Bernstein, Felix. 1932. *Die geographische Verteilung der Blutgruppen und ihre anthropologische Bedeutung*. Rome: Comitato Italiano per lo Studio dei Problemi della Popolazione / Istituto Poligrafico dello Stato.

Birchal, Telma S., and Sérgio D. J. Pena. 2011. The biological nonexistence versus the social existence of human races: Can science instruct the social ethos? In *Racial identities, genetic ancestry, and health in South America: Argentina, Brazil, Colombia, and Uruguay*, edited by Sahra Gibbon, Ricardo Ventura Santos, and Mónica Sans, 69–99. New York: Palgrave Macmillan.

Bisso Machado, Rafael. 2006. Negros, mas nem tão Africanos. Undergraduate thesis, Instituto de Biociências, Universidade Federal do Rio Grande do Sul, Porto Alegre.

Blanckaert, Claude. 1989. "L'anthropologie personnifiée": Paul Broca et la biologie du genre humain. In *Paul Broca: Mémoires d'anthropologie*, edited by Claude Blanckaert, i–xliii. Paris: Éditions Jean-Michel Place.

Bliss, Catherine. 2008. Mapping admixture by race. *International Journal of Technology, Knowledge and Society* 4(4): 79–83.

Bliss, Catherine. 2009a. Genome sampling and the biopolitics of race. In *A Foucault for the 21st century: Governmentality, biopolitics and discipline in the new millennium*, edited by Samuel Binkley and Jorge Capetillo, 322–339. Boston: Cambridge Scholars.

Bliss, Catherine. 2009b. The new science of race: Sociological analysis of the genomic debate over race. PhD diss., New School of Social Research, New York.

Bliss, Catherine. 2011. Racial taxonomy in genomics. *Social Science and Medicine* 73(7): 1019–1027.

Bolnick, Deborah A. 2008. Individual ancestry inference and the reification of race as a biological phenomenon. In *Revisiting race in a genomic age*, edited by Barbara A. Koenig, Sandra Soo-Jin Lee, and Sarah S. Richardson, 70–85. New Brunswick, NJ: Rutgers University Press.

Bolnick, Deborah A., Duana Fullwiley, Troy Duster, et al. 2007. The science and business of genetic ancestry testing. *Science* 318(5849): 399–400.

Bonfil Batalla, Guillermo. 1996. *México profundo: Reclaiming a civilization*. Translated by Philip A. Dennis. Austin: University of Texas Press.

Bonfil Batalla, Guillermo. 2004. Sobre la ideología del mestizaje. In *Decadencia y auge de identidades*, edited by José Valenzuela, 88–96. Mexico City: Ediciones Plaza y Valdés.

Bonil Gómez, Katherine. 2011. *Gobierno y calidad en el orden colonial: Las categorías del mestizaje en la provincia de Mariquita en la segunda mitad del siglo XVIII*. Bogotá: Universidad de los Andes.

Bonilla, C., B. Bertoni, S. Gonzalez, et al. 2004. Substantial Native American female contribution to the population of Tacuarembo, Uruguay, reveals past episodes of sex-biased gene flow. *American Journal of Human Biology* 16(3): 289–297.

Bornholdt, Luciano Campelo. 2010. What is a Gaúcho? Intersections between state, identities and domination in Southern Brazil. *(con)textos* 4: 23–41.

Bortolini, Maria Cátira. 2009. Comments on "Color, race, and genomic ancestry in Brazil," by R. V. Santos et al. *Current Anthropology* 50(6): 805.

Bortolini, Maria Cátira. 2012. Resposta ao trabalho de Kent e Santos: "Os charruas vivem" nos Gaúchos: A vida social de uma pesquisa de "resgate" genético de uma etnia indígena extinta no Sul do Brasil. *Horizontes Antropológicos* 18(37): 373–378.

Bortolini, Maria Cátira, Wilson Araújo Da Silva, Dinorah Castro De Guerra, et al. 1999. African-derived South American populations: A history of symmetrical and asymmetrical matings according to sex revealed by bi- and uni-parental genetic markers. *American Journal of Human Biology* 11(4): 551–563.

Bortolini, Maria Cátira, Tania De Azevedo Weimer, Francisco M. Salzano, et al. 1995. Evolutionary relationships between black South American and African populations. *Human Biology* 67: 547.

Bortolini, Maria Cátira, Mark G. Thomas, Lounes Chikhi, et al. 2004. Ribeiro's typology, genomes, and Spanish colonialism, as viewed from Gran Canaria and Colombia. *Genetics and Molecular Biology* 27(1): 1–8.

Bortolini, Maria Cátira, M. A. Zago, F. M. Salzano, et al. 1997. Evolutionary and anthropological implications of mitochondrial DNA variation in African Brazilian populations. *Human Biology* 69(2): 141–159.

Bourdieu, Pierre, and Loïc Wacquant. 1999. On the cunning of imperialist reason. *Theory, Culture and Society* 16(1): 41–58.

Bowker, Geoffrey C., and Susan Leigh Star. 1999. *Sorting things out: Classification and its consequences.* Cambridge, MA: MIT Press.

Bowler, Peter. J. 1989. *Evolution: The history of an idea.* Berkeley: University of California Press.

Bracco, Diego. 2004. *Charrúas, guenoas y guaraníes: Interacción y destrucción: Indígenas en el Río de la Plata.* Montevideo: Linardi y Risso.

Braun, Lundy, Anne Fausto-Sterling, Duana Fullwiley, et al. 2007. Racial categories in medical practice: How useful are they? *PLoS Medicine* 4(9): e271.

Briceño, Ignacio. 1990. La Expedición Humana en el Chocó. *Boletín Expedición Humana 1992*, no. 4: 2.

Brodwin, Paul. 2002. Genetics, identity and the anthropology of essentialism. *Anthropological Quarterly* 75: 323–330.

Brodwin, Paul. 2005. "Bioethics in action" and human population genetics research. *Culture, Medicine and Psychiatry* 29(2): 145–178.

Brown, Ryan A., and George J. Armelagos. 2001. Apportionment of racial diversity: A review. *Evolutionary Anthropology* 10: 34–40.

Builes, J. J., N. Alzate, C. Espinal, et al. 2008. Analysis of 16 Y-chromosomal STRs in an African descent sample population of Chocó (Colombia). *Forensic Science International: Genetics Supplement Series* 1(1): 184–186.

Builes, J. J., M. L. J. Bravo, A. Montoya, et al. 2004. Population genetics of eight new Y-chromosomal STR haplotypes in three Colombian populations: Antioquia, Chocó and Cartagena. *International Congress Series* 1261(0): 310–312.

Burchard, Esteban Gonzalez, Luisa N. Borrell, Shweta Choudhry, et al. 2005. Latino populations: A unique opportunity for the study of race, genetics, and social environment in epidemiological research. *American Journal of Public Health* 95(12): 2161–2168.

Burchard, Esteban Gonzalez, Elad Ziv, Natasha Coyle, et al. 2003. The importance of race and ethnic background in biomedical research and clinical practice. *New England Journal of Medicine* 348(12): 1170–1175.

Bustamante, Carlos D., Francisco M. De La Vega, and Esteban G. Burchard. 2011. Genomics for the world. *Nature* 475(7355): 163–165.

Canizales-Quinteros, Samuel. 2011. Genética de la obesidad en la población mexicana. Paper presented at the INMEGEN-Nestlé International Symposium on Nutrigenomics, Mexico City, 29–30 September.

Carrillo, Ana María. 2001. Los médicos y la "degeneración de la raza indígena." *Ciencias* 60–61: 64–71.

Carvajal-Carmona, Luis G., Roel Ophoff, Susan Service, et al. 2003. Genetic demography of Antioquia (Colombia) and the Central Valley of Costa Rica. *Human Genetics* 112(5): 534–541.

Carvajal-Carmona, Luis G., Iván D. Soto, Nicolás Pineda, et al. 2000. Strong Amerind/white sex bias and a possible Sephardic contribution among the founders of a population in northwest Colombia. *American Journal of Human Genetics* 67(5): 1287–1295.

Carvalho-Silva, Denise, Fabrício R. Santos, Jorge Rocha, et al. 2001. The phylogeography of Brazilian Y-chromosome lineages. *American Journal of Human Genetics* 68: 281–286.

Castro-Faria, Luiz de. 1952. Pesquisas de antropologia física no Brasil. *Boletim do Museu Nacional* 13: 1–106.

Castro-Faria, Luiz de. 1993. *Antropologia: Espetáculo e excelência.* Rio de Janeiro: Editora UFRJ.

Castro-Santos, Luiz Antônio de. 1985. O pensamento sanitarista na Primeira República: Uma ideologia de construção da nacionalidade. *Dados—Revista de Ciências Sociais* 28(2): 193–210.

Caulfield, Sueann. 2000. *In defense of honor: Sexual morality, modernity, and nation in early-twentieth-century Brazil.* Durham, NC: Duke University Press.

Cavalli-Sforza, L. Luca, Paolo Menozzi, and Alberto Piazza. 1994. *The history and geography of human genes.* Princeton, NJ: Princeton University Press.

Cerda-Flores, Ricardo M., Maria C. Villalobos-Torres, Hugo A. Barrera-Saldaña, et al. 2002. Genetic admixture in three Mexican Mestizo populations based on D1S80 and HLA-DQA1 loci. *American Journal of Human Biology* 14(2): 257–263.

Chadarevian, Soraya de. 2002. *Designs for life: Molecular biology after World War II.* Cambridge: Cambridge University Press.

Chadwick, Ruth, and Kare Berg. 2001. Solidarity and equity: New ethical frameworks for genetic databases. *Nature Reviews Genetics* 2(4): 318–321.

Chakraborty, Ranajit, and Kenneth M. Weiss. 1988. Admixture as a tool for finding linked genes and detecting that difference from allelic association between loci. *Proceedings of the National Academy of Sciences* 85(23): 9119–9123.

Chamberlin, J. Edward, and Sander L. Gilman, eds. 1985. *Degeneration: The dark side of progress.* New York: Columbia University Press.

Chong, Natividad Gutiérrez. 2008. Symbolic violence and sexualities in the myth making of Mexican national identity. *Ethnic and Racial Studies* 31(3): 524–542.

Choudhry, Shweta, Natasha Coyle, Hua Tang, et al. 2006. Population stratification confounds genetic association studies among Latinos. *Human Genetics* 118(5): 652–664.

CINEP. 1998. *Colombia país de regiones.* Bogotá: CINEP.

Coimbra, Carlos E. A., Nancy M. Flowers, Francisco M. Salzano, et al. 2004. *The Xavante in transition: Health, ecology, and bioanthropology in Central Brazil*. Ann Arbor: University of Michigan Press.

Comaroff, John L., and Jean Comaroff. 2009. *Ethnicity, Inc.* Chicago: University of Chicago Press.

Condit, Celeste Michelle. 1999. *The meanings of the gene: Public debates about human heredity*. Madison: University of Wisconsin Press.

Condit, Celeste Michelle, Roxanne Parrott, and Tina M. Harris. 2002. Lay understandings of the relationship between race and genetics: Development of a collectivized knowledge through shared discourse. *Public Understanding of Science* 11: 373–387.

Condit, Celeste M., Roxanne L. Parrott, Tina M. Harris, et al. 2004. The role of "genetics" in popular understandings of race in the United States. *Public Understanding of Science* 13(3): 249–272.

Contini, Gianfranco. 1959. I più antichi esempî di «razza». *Studi di filologia italiana, Bollettino annuale dell'Accademia della Crusca, Firenze* 17: 319–327.

Contreras, Alejandra V., Tulia Monge-Cazares, Luis Alfaro-Ruiz, et al. 2011. Resequencing, haplotype construction and identification of novel variants of CYP2D6 in Mexican Mestizos. *Pharmacogenomics* 12(5): 745–756.

Cooper, Richard S., Jay S. Kaufman, and Ryk Ward. 2003. Race and genomics. *New England Journal of Medicine* 348(12): 1166–1170.

Corrêa, Mariza. 1982. As ilusões da liberdade: A escola Nina Rodrigues e a antropologia no Brasil. PhD diss., Universidade de São Paulo.

Corrigan, Oonagh, and Richard Tutton, eds. 2004. *Genetic databases: Socio-ethical issues in the collection and use of* dna. London: Routledge.

Cunha, Olívia Maria Gomes. 2002. *Intenção e gesto: Pessoa, cor e a produção cotidiana da (in)diferença no Rio de Janeiro (1927–1942)*. Rio de Janeiro: Arquivo Nacional.

Cunningham, Hilary. 1998. Colonial encounters in postcolonial contexts: Patenting Indigenous DNA and the Human Genome Diversity Project. *Critique of Anthropology* 18(2): 205–233.

Dacanal, José Hildebrando. 1980. A miscigenação que não houve. In rs: *Cultural e ideologia*, edited by José Hildebrando Dacanal and Sergius Gonzaga, 25–33. Porto Alegre: Mercado Aberto.

Darvasi, Ariel, and Sagiv Shifman. 2005. The beauty of admixture. *Nature Genetics* 37(2): 118–119.

De la Cadena, Marisol. 2000. *Indigenous mestizos: The politics of race and culture in Cuzco, Peru, 1919–1991*. Durham, NC: Duke University Press.

De la Peña, Guillermo. 2006. A new Mexican nationalism? Indigenous rights, constitutional reform and the conflicting meanings of multiculturalism. *Nations and Nationalism* 12(2): 279–302.

Deleuze, Gilles. 1992. Postscript on society of control. *October* 59: 3–7.

Deleuze, Gilles. 2006. *Foucault*. Translated by Seán Hand. London: Continuum.

Dennis, Carina. 2003. Special section on human genetics: The rough guide to the genome. *Nature* 425(6960): 758–759.

Diacon, Todd A. 2004. *Stringing together a nation: Cândido Mariano da Silva Rondon and the construction of a modern Brazil, 1906–1930*. Durham, NC: Duke University Press.

Diamond, Stanley. 1993. *In search of the primitive: A critique of civilization*. New Brunswick, NJ: Transaction.

Díaz, Luisa Fernanda. 2010. Análisis de 17 loci de STR de cromosoma Y en las poblaciones de Bogotá y Santander con fines genético poblacionales y forenses. MA thesis, Universidad Javeriana, Bogotá.

Díaz del Castillo H., Adriana, María Fernanda Olarte Sierra, and Tania Pérez-Bustos. 2012. Testigos modestos y poblaciones invisibles en la cobertura de la genética humana en los medios de comunicación colombianos. *Interface: Comunicação, Saúde, Educação* 16(41): 451–467.

Dornelles, C. L., S. M. Callegari-Jacques, W. M. Robinson, et al. 1999. Genetics, surnames, grandparents' nationalities, and ethnic admixture in southern Brazil: Do the patterns of variation coincide? *Genetics and Molecular Biology* 22(2): 151–161.

Duster, Troy. 2003a. *Backdoor to eugenics*. 2nd ed. London: Routledge.

Duster, Troy. 2003b. Buried alive: The concept of race in science. In *Genetic nature/culture: anthropology and science beyond the two-culture divide*, edited by Alan H. Goodman, Deborah Heath, and Susan M. Lindee, 258–277. Berkeley: University of California Press.

Duster, Troy. 2006a. Comparative perspectives and competing explanations: Taking on the newly configured reductionist challenge to sociology. *American Sociological Review* 71(1): 1–15.

Duster, Troy. 2006b. The molecular reinscription of race: Unanticipated issues in biotechnology and forensic science. *Patterns of Prejudice* 40(4–5): 427–441.

Edwards, Jeanette, and Carles Salazar, eds. 2009. *European kinship in the age of biotechnology*. Oxford: Berghahn.

Ellison, George T. H., Andrew Smart, Richard Tutton, et al. 2007. Racial categories in medicine: A failure of evidence-based practice? *PLoS Medicine* 4(9): e287.

Epstein, Steven. 2007. *Inclusion: The politics of difference in medical research*. Chicago: University of Chicago Press.

Escobar, Arturo. 2008. *Territories of difference: Place, movements, life, redes*. Durham, NC: Duke University Press.

Fabian, Johannes. 2002. *Time and the other: How anthropology makes its object*. New York: Columbia University Press.

Falcón, Romana. 1996. *Españoles y mexicanos a mediados del siglo XIX*. Mexico City: El Colegio de México.

Fausto-Sterling, Anne. 2000. *Sexing the body: Gender politics and the construction of sexuality*. New York: Basic Books.

Fausto-Sterling, Anne. 2004. Refashioning race: DNA and the politics of health care. *differences: A Journal of Feminist Cultural Studies* 15(3): 1–37.

Fawcet de Posada, Louise, and Eduardo Posada Carbó. 1992. En la tierra de las oportunidades: Los sirio-libaneses en Colombia. *Boletín Cultural y Bibliográfico* 29: 3–22.

Ferreira, Luiz Otávio. 2008. O ethos positivista e a institucionalização das ciências no Brasil. In *Antropologia brasiliana: Ciência e educação na obra de Edgard Roquette-Pinto*, edited by Nísia Trindade Lima and Dominichi Miranda de Sá, 87–98. Rio de Janeiro: Editora U F M G and Editora Fiocruz.

Fletcher, Ian Christopher. 2005. Introduction: New historical perspectives on the First Universal Races Congress of 1911. *Radical History Review* 92: 99–102.

Fog, Lisbeth. 2006. Emilio Yunis Turbay, perfiles de personajes científicos destacados. Universia. Accessed 5 April 2013. http://especiales.universia.net.co/galeria-de-cientificos/ciencias-de-la-salud/emilio-yunis-turbay.html.

Forbes, Jack D. 1993. *Africans and Native Americans: The language of race and the evolution of red-black peoples*. Urbana: University of Illinois Press.

Foucault, Michel. 2008. *The birth of biopolitics: Lectures at the Collège de France, 1978–1979*. Translated by Graham Burchell. Edited by Michel Senellart, Arnold I. Davidson, Alessandro Fontana, and Francois Ewald. New York: Palgrave Macmillan.

Foucault, Michel, Mauro Bertani, Alessandro Fontana, et al. 2003. *Society must be defended: Lectures at the Collège de France, 1975–76*. London: Picador.

Franco, M. Helena L. P., Tania A. Weimer, and Francisco M. Salzano. 1982. Blood polymorphisms and racial admixture in two Brazilian populations. *American Journal of Physical Anthropology* 58(2): 127–132.

Franklin, Sarah. 2001. Biologization revisited: Kinship theory in the context of the new biologies. In *Relative values: Reconfiguring kinship studies*, edited by Sarah Franklin and Susan McKinnon, 303–325. Durham, NC: Duke University Press.

Freire-Maia, Newton. 1973. *Brasil: Laboratório racial*. Petrópolis: Vozes.

French, Jan Hoffman. 2009. *Legalizing identities: Becoming black or Indian in Brazil's northeast*. Chapel Hill: University of North Carolina Press.

Freyre, Gilberto. 1946 [1933]. *Casa-grande e senzala: Formação da família brasileira sob o regime de economia patriarcal*. Rio de Janeiro: José Olympio.

Friedemann, Nina S. de. 1984. Estudios de negros en la antropología colombiana. In *Un siglo de investigación social: Antropología en Colombia*, edited by Jaime Arocha and Nina de Friedemann, 507–572. Bogotá: Etno.

Friedemann, Nina S. de. 1990. La América negra y también oculta: Perfiles etnomédicos y genéticos en el Litoral Pacífico. *Boletín Expedición Humana 1992*, no. 5: 1.

Friedemann, Nina S. de, and Diógenes Fajardo. 1993. La herencia de Caín: Entrevista con el médico genetista Jaime Bernal Villegas. *América Negra* 5: 207–215.

Fry, Peter. 2000. Politics, nationality, and the meanings of "race" in Brazil. *Daedalus* 129(2): 83–118.

Fry, Peter. 2005a. *A persistência da raça: Ensaios antropológicos sobre o Brasil e a África austral*. Rio de Janeiro: Civilização Brasileira.

Fry, Peter. 2005b. O significado da anemia falciforme no contexto da "política racial" do governo brasileiro 1995–2004. *História, Ciências, Saúde-Manguinhos* 12(2): 374–370.

Fry, Peter, Yvonne Maggie, Marcos Chor Maio et al., eds. 2007. *Divisões perigosas: Políticas raciais no Brasil contemporâneo.* Rio de Janeiro: Civilização Brasileira.

Fujimura, Joan H., Troy Duster, and Ramya Rajagopalan. 2008. Introduction: Race, genetics, and disease: Questions of evidence, matters of consequence. *Social Studies of Science* 38(5): 643–656.

Fujimura, Joan H., and Ramya Rajagopalan. 2011. Different differences: The use of "genetic ancestry" versus race in biomedical human genetic research. *Social Studies of Science* 41(1): 5–30.

Fullwiley, Duana. 2007a. The molecularization of race: Institutionalizing human difference in pharmacogenetics practice. *Science as Culture* 16(1): 1–30.

Fullwiley, Duana. 2007b. Race and genetics: Attempts to define the relationship. *BioSocieties* 2(2): 221–237.

Fullwiley, Duana. 2008. The biologistical construction of race: "Admixture" technology and the new genetic medicine. *Social Studies of Science* 38(5): 695–735.

Galanter, Joshua Mark, Juan Carlos Fernandez-Lopez, Christopher R. Gignoux, et al. 2012. Development of a panel of genome-wide ancestry informative markers to study admixture throughout the Americas. *PLoS Genetics* 8(3): e1002554.

Gall, Olivia. 2004. Identidad, exclusión y racismo: Reflexiones teóricas y sobre México. *Revista Mexicana de Sociología* 66(2): 221–259.

Gall, Olivia. 2007. *Racismo, mestizaje y modernidad: Visiones desde latitudes diversas.* Mexico City: UNAM.

Gannett, Lisa. 2004. The biological reification of race. *British Journal for the Philosophy of Science* 55(2): 323–345.

García Deister, Vivette. 2011. Mestizaje en el laboratorio, una toma instantánea. In *Genes (y) mestizos: Genómica y raza en la biomedicina mexicana,* edited by Carlos López Beltrán, 143–154. Mexico City: Ficticia Editorial.

Garrido, Margarita. 2005. "Free men of all colours" in New Granada: Identity and obedience before Independence. In *Political cultures in the Andes, 1750–1950,* edited by Cristóbal Aljovín de Losada and Nils Jacobsen, 165–183. Durham, NC: Duke University Press.

Gaspar Neto, Verlan Valle, and Ricardo Ventura Santos. 2011. Biorrevelações: Testes de ancestralidade genética em perspectiva antropológica comparada. *Horizontes Antropológicos* 35: 227–255.

Gaspar Neto, Verlan Valle, Ricardo Ventura Santos, and Michael Kent. 2012. Biorrevelações: Testes de ancestralidade genética em perspectiva antropológica comparada. In *Identidades emergentes, genética e saúde: Perspectivas antropológicas,* edited by Ricardo Ventura Santos, Sahra Gibbon, and Jane F. Beltrão, 233–267. Rio de Janeiro: Editora Garamond and Editora Fiocruz.

Gaspari, Elio. 2000. O branco tem a marca de Nana. *Folha de São Paulo,* 16 April, A14.

Genebase Tutorials. 2013. Learn about Y-DNA Haplogroup Q. Accessed 26 June. http://www.genebase.com/learning/article/16.

Gibbon, Sahra, and Carlos Novas, eds. 2007. *Biosocialities, genetics and the social sciences: Making biologies and identities.* London: Routledge.

Gibbon, Sahra, Ricardo Ventura Santos, and Mónica Sans, eds. 2011. *Racial identities, genetic ancestry, and health in South America: Argentina, Brazil, Colombia, and Uruguay.* New York: Palgrave Macmillan.

Gieryn, Thomas F. 1983. Boundary-work and the demarcation of science from non-science: Strains and interests in professional ideologies of scientists. *American Sociological Review* 48(6): 781–795.

Gilroy, Paul. 2000. *Against race: Imagining political culture beyond the color line.* Cambridge, MA: Harvard University Press.

Gingrich, André, and Richard G. Fox, eds. 2002. *Anthropology, by comparison.* New York: Routledge.

Gómez, Alberto. 1991. Entre los embera-epena. *Boletín Expedición Humana 1992,* no. 10: 8.

Gómez, Alberto. 1992. El Banco Biológico Humano. *Revista Javeriana* 118(586): 9–11.

Gómez Gutiérrez, Alberto. 1998. *Al cabo de las velas: Expediciones científicas en Colombia, S. XVIII, XIX y XX.* Bogotá: Instituto Colombiano de Cultura Hispánica.

Gómez Gutiérrez, Alberto. 2011. Entrevistas con científicos galardonados. Premio de Ciencias Exactas, Físicas y Naturales, Fundación Álejandro Ángel Escobar. Accessed 5 April 2013. http://www.faae.org.co/html/ganadoresanoc.htm.

Gómez Gutiérrez, Alberto, Ignacio Briceño Balcázar, and Jaime Eduardo Bernal Villegas. 2011. Patrones de identidad genética en poblaciones contemporáneas y precolombinas. Fundación Alejandro Ángel Escobar. Accessed 5 April 2013. http://www.faae.org.co/html/resena/2011-identidad-genetica .html?keepThis=true&TB_iframe=true&height=380&width=628.

Gómez Izquierdo, Jorge José. 2005. Racismo y nacionalismo en el discurso de las élites mexicanas. In *Caminos del racismo en México,* edited by Jorge José Gómez Izquierdo, 117–181. Mexico City: Plaza y Valdés.

Gonçalves, V. F., F. Prosdocimi, L. S. Santos, et al. 2007. Sex-biased gene flow in African Americans but not in American Caucasians. *Genetics and Molecular Research* 6(2): 156–161.

González Burchard, Esteban, Luisa N. Borrell, Shweta Choudhry, et al. 2005. Latino populations: A unique opportunity for the study of race, genetics, and social environment in epidemiological research. *American Journal of Public Health* 95(12): 2161–2168.

Goodman, Alan H., Deborah Heath, and M. Susan Lindee, eds. 2003. *Genetic nature/ culture: Anthropology and science beyond the two-culture divide.* Berkeley: University of California Press.

Gorodezky, C., L. Terán, and A. Escobar Gutiérrez. 1979. HLA frequencies in a Mexican Mestizo population. *Tissue Antigens* 14(4): 347–352.

Gotkowitz, Laura, ed. 2011a. *Histories of race and racism: The Andes and Mesoamerica from colonial times to the present.* Durham, NC: Duke University Press.

Gotkowitz, Laura. 2011b. Introduction: Racisms of the present and the past in Latin America. In *Histories of race and racism: The Andes and Mesoamerica from colonial times to the present,* edited by Laura Gotkowitz, 1–53. Durham, NC: Duke University Press.

Gould, Stephen Jay. 1996. *The mismeasure of man.* New York: Norton.

Graham, Richard, ed. 1990. *The idea of race in Latin America, 1870–1940.* Austin: University of Texas Press.

Gray, Ann. 2002. *Research practice for cultural studies: Ethnographic methods and lived cultures.* London: Sage.

Green, Lance D., James N. Derr, and Alec Knight. 2000. mtDNA affinities of the peoples of North-Central Mexico. *American Journal of Human Genetics* 66(3): 989–998.

Grin, Monica. 2010. "Raça": *Debate público no Brasil.* Rio de Janeiro: Editora Mauad/FAPERJ.

Gros, Christian. 1991. *Colombia indígena: Identidad cultural y cambio social.* Bogotá: CEREC.

Guardado-Estrada, Mariano, Eligia Juarez-Torres, Ingrid Medina-Martinez, et al. 2009. A great diversity of Amerindian mitochondrial DNA ancestry is present in the Mexican mestizo population. *Journal of Human Genetics* 54(12): 695–705.

Guerreiro, João Farias, Ândrea Kely Campos Ribeiro-dos-Santos, Eduardo José Melo dos Santos, et al. 1999. Genetical-demographic data from two Amazonian populations composed of descendants of African slaves: Pacoval and Curiau. *Genetics and Molecular Biology* 22(2): 163–167.

Guimarães, Antonio Sérgio. 1997. A desigualdade que anula a desiguladade, notas sobre a ação afirmativa no Brasil. In *Multiculturalismo e racismo: Una comparação Brasil–Estados Unidos,* edited by Alayde Sant'Anna and Jessé Souza, 233–242. Brasilia: Paralelo 15.

Guimarães, Antonio Sérgio. 1999. *Racismo e anti-racismo no Brasil.* Rio de Janeiro: Editora 34.

Guimarães, Antonio Sérgio. 2007. Racial democracy. In *Imagining Brazil,* edited by Jessé Souza and Valter Sinder, 119–140. Lanham, MD: Lexington.

Hale, Charles R. 2005. Neoliberal multiculturalism: The remaking of cultural rights and racial dominance in Central America. *PoLAR: Political and Legal Anthropology Review* 28(1): 10–28.

Haraway, Donna. 1988. Situated knowledges: The science question in feminism and the privilege of partial perspective. *Feminist Studies* 14(3): 575–599.

Haraway, Donna. 1989. *Primate visions: Gender, race and nature in the world of modern science.* New York: Routledge.

Haraway, Donna. 1997. *Modest_Witness@Second_Millenium.FemaleMan©_Meets_Oncomouse™.* London: Routledge.

Hardy, John, and Andrew Singleton. 2009. Genomewide association studies and human disease. *New England Journal of Medicine* 360(17): 1759–1768.

Hartigan, John. 2013a. Looking for race in the Mexican Book of Life: INMEGEN and the Mexican Genome Project. In *Anthropology of race: Genes, biology, and culture,* edited by John Hartigan, 125–150. Santa Fe, NM: School for Advanced Research Press.

Hartigan, John. 2013b. Translating "race" and "raza" between the United States and Mexico. *North American Dialogue* 16(1): 29–41.

Harvey, Joy Dorothy. 1983. *Races specified, evolution transformed: The social context of scientific debates originating in the Société d'Anthropologie de Paris.* Cambridge, MA: Harvard University Press.

Heath, Deborah, Rayna Rapp, and Karen-Sue Taussig. 2007. Genetic citizenship. In *A companion to the anthropology of politics*, edited by David Nugent and Joan Vincent, 152–167. New York: Blackwell.

Hedgecoe, Adam. 1998. Geneticization, medicalisation and polemics. *Medicine, Health Care and Philosophy* 1(3): 235–243.

Hering Torres, Max Sebastián. 2003. "Limpieza de sangre": ¿Racismo en la edad moderna? *Tiempos Modernos* 9: 1–16.

Hey, Tony, Stewart Tansley, and Kristin M. Tolle, eds. 2009. *The fourth paradigm: Data-intensive scientific discovery*. Redmond, WA: Microsoft Research.

Hindorff, Lucia A., Praveen Sethupathy, Heather A. Junkins, et al. 2009. Potential etiologic and functional implications of genome-wide association loci for human diseases and traits. *Proceedings of the National Academy of Sciences* 106(23): 9362–9367.

Hirschhorn, Joel N., and Mark J. Daly. 2005. Genome-wide association studies for common diseases and complex traits. *Nature Reviews Genetics* 6(2): 95–108.

Hobsbawm, Eric, and Terence Ranger, eds. 1983. *The invention of tradition*. Cambridge: Cambridge University Press.

Hodgen, Margaret T. 1964. *Early anthropology in the sixteenth and seventeenth centuries*. Philadelphia: University of Pennsylvania Press.

Hoffmann, Odile. 2006. Negros y afromestizos en México: Viejas y nuevas lecturas de un mundo olvidado. *Revista Mexicana de Sociología* 68(1): 103–135.

Hoffmann, Odile, and María Teresa Rodríguez, eds. 2007. *Retos de la diferencia: Los actores de la multiculturalidad entre México y Colombia*. Mexico City: Centro de Estudios Mexicanos y Centroamericanos.

Houot, Annie. 2002. *Un cacique Charrúa en París*. Montevideo: Editorial Costa Atlantica.

Htun, Mala. 2004. From "racial democracy" to affirmative action: Changing state policy on race in Brazil. *Latin American Research Review* 39(1): 60–89.

Hunemeier, T., C. Carvalho, A. R. Marrero, et al. 2007. Niger-Congo speaking populations and the formation of the Brazilian gene pool: mtDNA and Y-chromosome data. *American Journal of Physical Anthropology* 133(2): 854–867.

Inda, Jonathan Xavier, and Renato Rosaldo. 2002. *The anthropology of globalization: A reader*. Oxford: Blackwell.

International HapMap Consortium. 2004. Integrating ethics and science in the International HapMap Project. *Nature Reviews Genetics* 5(6): 467–475.

Izquierdo, Jorge, and María Eugenia Sánchez Díaz de Rivera. 2011. *La ideología mestizante, el guadalupanismo y sus repercusiones sociales. Una revisión crítica de la "identidad nacional."* Puebla: Universidad Iberoamericana de Puebla.

Jaramillo Uribe, Jaime. 1968. *La sociedad neogranadina*. Vol. 1, *Ensayos sobre historia social colombiana*. Bogotá: Universidad Nacional de Colombia.

Jardine, Nicholas, James A. Secord, and Emma C. Spary, eds. 1996. *Cultures of natural history*. Cambridge: Cambridge University Press.

Jasanoff, Sheila. 2004a. Ordering knowledge, ordering society. In *States of knowledge:*

The co-production of science and social order, edited by Sheila Jasanoff, 13–45. London: Routledge.

Jasanoff, Sheila, ed. 2004b. States of knowledge: The co-production of science and social order. London: Routledge.

Jiménez López, Miguel, Luis López de Mesa, Calixto Torres Umaña, et al. 1920. Los problemas de la raza en Colombia. Bogotá: El Espectador.

Jiménez-Sánchez, Gerardo. 2009. Mapa del genoma de los mexicanos. Resumen ejecutivo 2009. Internal report. Mexico City: INMEGEN.

Jiménez-Sánchez, Gerardo, Barton Childs, and David Valle. 2001. Human disease genes. Nature 409(6822): 853–855.

Johnson, Lyman L., and Sonya Lipsett-Rivera, eds. 1998. The faces of honor: Sex, shame, and violence in colonial Latin America. Albuquerque: University of New Mexico Press.

Kahn, Jonathan. 2005. From disparity to difference: How race-specific medicines may undermine policies to address inequalities in health care. Southern California Interdisciplinary Law Journal 15(1): 105–129.

Kahn, Jonathan. 2008. Exploiting race in drug development: BiDil's interim model of pharmacogenomics. Social Studies of Science 38(5): 737–758.

Kaplan, Judith B., and Trude Bennett. 2003. Use of race and ethnicity in biomedical publication. jama 289(20): 2709–2716.

Karafet, Tatiana M., Fernando L. Mendez, Monica B. Meilerman, et al. 2008. New binary polymorphisms reshape and increase resolution of the human Y chromosomal haplogroup tree. Genome Research 18(5): 830–838.

Katzew, Ilona. 2005. Casta painting: Images of race in eighteenth-century Mexico. New Haven, CT: Yale University Press.

Katzew, Ilona, and Susan Deans-Smith, eds. 2009. Race and classification: The case of Mexican America. Stanford, CA: Stanford University Press.

Keller, Evelyn Fox. 1995. Refiguring life: Metaphors of twentieth-century biology. New York: Columbia University Press.

Kent, Michael. 2013. The importance of being Uros: Indigenous identity politics in the genomic age. Social Studies of Science 43(4): 534–556.

Kent, Michael, and Ricardo Ventura Santos. 2012a. Genes, boleadeiras e abismos colossais: Elementos para um diálogo entre genética e antropologia. Horizontes Antropológicos 18(37): 379–384.

Kent, Michael, and Ricardo Ventura Santos. 2012b. "Os charruas vivem" nos Gaúchos: A vida social de uma pesquisa de "resgate" genético de uma etnia indígena extinta no Sul do Brasil. Horizontes Antropológicos 18(37): 341–372.

Kent, Michael, Ricardo Ventura Santos, and Peter Wade. 2012. Negotiating imagined genetic communities in Brazil: Tensions between narratives of unity and diversity. Unpublished manuscript.

Knight, Alan. 1990. Racism, revolution and indigenismo in Mexico, 1910–1940. In The idea of race in Latin America, edited by Richard Graham, 71–113. Austin: University of Texas Press.

Koenig, Barbara A., Sandra Soo-Jin Lee, and Sarah S. Richardson, eds. 2008. *Revisiting race in a genomic age*. New Brunswick, NJ: Rutgers University Press.

Krieger, H., N. E. Morton, M. P. Mi, et al. 1965. Racial admixture in north-eastern Brazil. *Annals of Human Genetics* 29(2): 113–125.

Kropf, Simone P., Nara Azevedo, and Luiz Otávio Ferreira. 2003. Biomedical research and public health in Brazil: The case of Chagas' disease (1909–1950). *Social History of Medicine* 6: 111–129.

Kuper, Adam. 1988. *The invention of primitive society: Transformation of an illusion*. London: Routledge.

Laboratorio de Genética Humana. 2013. Laboratorio de Genética Humana, 1978–2012. Accessed 26 July. http://geneticahumana.uniandes.edu.co/Laboratorio_Genetica_Humana/Bienvenida.html.

Lacerda, João Baptista de. 1875. Documents pour servir à l'histoire de l'homme fossile du Brésil. *Mémoires de la Societé d'Anthropologie de Paris* 2: 517–542.

Lacerda, João Baptista de. 1876. Contribuições para o estudo anthropologico das raças indigenas do Brasil: Nota sobre a conformação dos dentes. *Archivos do Museu Nacional* 1: 77–83.

Lacerda, João Baptista de. 1882a. Botocudos. In *Revista da Exposição Anthropologica Brasileira*, edited by Mello Moraes Filho. Rio de Janeiro: Typographia de Pinheiro.

Lacerda, João Baptista de. 1882b. A força muscular e a delicadeza dos sentidos nos nosso indigenas. In *Revista da Exposição Anthropologica Brasileira*, edited by Mello Moraes Filho. Rio de Janeiro: Typographia de Pinheiro.

Lacerda, João Baptista de. 1882c. A morphologia craneana do homem dos sambaquis. In *Revista da Exposição Anthropologica Brasileira*, edited by Mello Moraes Filho. Rio de Janeiro: Typographia de Pinheiro.

Lacerda, João Baptista de. 1905. *Fastos do Museu Nacional do Rio de Janeiro: Recordações historicas e scientificas fundadas em documentos authenticos e informações veridicas*. Rio de Janeiro: Imprensa Nacional.

Lacerda, João Baptista de. 1911. *Sur les métis au Brésil*. Paris: Imprimerie Devouge.

Lacerda, João Baptista de, and José Rodrigues Peixoto. 1876. Contribuições para o estudo anthropologico das raças indigenas do Brasil. *Archivos do Museu Nacional* 1: 47–75.

Lara, José Luis Trueba. 1990. *Los chinos en Sonora: Una historia olvidada*. Hermosillo: Instituto de Investigaciones Históricas, Universidad de Sonora.

Lasker, Gabriel Ward. 1954. Photoelectric measurement of skin color in a Mexican Mestizo population. *American Journal of Physical Anthropology* 12(1): 115–121.

Latour, Bruno. 1987. *Science in action: How to follow scientists and engineers through society*. Milton Keynes, U.K.: Open University Press.

Latour, Bruno. 1990. Visualisation and cognition: Drawing things together. In *Representation in scientific practice*, edited by Michael Lynch and Steve Woolgar, 19–68. Cambridge, MA: MIT Press.

Latour, Bruno. 1993. *We have never been modern*. Cambridge, MA: Harvard University Press.

Latour, Bruno. 1999. *Pandora's hope: Essays on the reality of science studies*. Cambridge, MA: Harvard University Press.

Latour, Bruno. 2004. Why has critique run out of steam? From matters of fact to matters of concern. *Critical Inquiry* 30(2): 225–248.

Latour, Bruno. 2005. *Reassembling the social: An introduction to actor-network theory*. Clarendon Lectures in Management Studies. Oxford: Oxford University Press.

Latour, Bruno, and Steve Woolgar. 1986. *Laboratory life: The construction of scientific facts*. 2nd ed. Princeton, NJ: Princeton University Press.

Law, John. 1991. Introduction: Monsters, machines and sociotechnical relations. In *A sociology of monsters: Essays on power, technology and dominations*, edited by John Law, 1–23. London: Routledge.

Law, John. 2008. Actor-network theory and material semiotics. In *The new Blackwell companion to social theory*, edited by Bryan S. Turner, 141–158. Oxford: Blackwell.

Leal León, Claudia. 2010. Usos del concepto "raza" en Colombia. In *Debates sobre ciudadanía y políticas raciales en las Américas Negras*, edited by Claudia Mosquera Rosero-Labbé, Agustín Laó Montes, and César Rodríguez Garavito, 389–438. Bogotá: Universidad Nacional de Colombia.

Leite, Fabio P. N., Sidney E. B. Santos, Elzemar M. R. Rodríguez, et al. 2009. Linkage disequilibrium patterns and genetic structure of Amerindian and non-Amerindian Brazilian populations revealed by long-range X-STR markers. *American Journal of Physical Anthropology* 139(3): 404–412.

Leite, Ilka Boaventura. 1996. *Negros no sul do Brasil: Invisibilidade e territorialidade*. Florianopolis: Letras Contemporâneas.

Leonelli, Sabina. 2010. Packaging small facts for re-use: Databases in model organism biology. In *How well do facts travel?*, edited by Peter Howlett and Mary S. Morgan, 325–348. New York: Cambridge University Press.

Leonelli, Sabina, and Rachel A. Ankeny. 2012. Re-thinking organisms: The impact of databases on model organism biology. *Studies in History and Philosophy of Biological and Biomedical Sciences* 43(1): 29–36.

Lewontin, Richard C. 1972. The apportionment of human diversity. *Evolutionary Biology* 6: 381–398.

Lewontin, Richard C., Steven Rose, and Leon Kamin. 1984. *Not in our genes: Biology, ideology and human nature*. New York: Pantheon.

Liberman, Anatoly. 2009. The Oxford etymologist looks at race, class and sex. Accessed 27 August 2011. OUPblog. http://blog.oup.com/2009/04/race-2/.

Lima, Nísia Trindade. 2007. Public health and social ideas in modern Brazil. *American Journal of Public Health* 97: 1209–1215.

Lima, Nísia Trindade, and Nara Britto. 1996. Salud y nación: Propuesta para el saneamiento rural. Un estudio de la *Revista Saúde* (1918–1919). In *Salud, cultura y sociedad en America Latina*, edited by Marcos Cueto, 135–158. Lima: Instituto de Estudos Peruanos, Organización Panamericana de la Salud.

Lima, Nísia Trindade, and Gilberto Hochman. 1996. Condenado pela raça, absolvido pela medicina: O Brasil descoberto pelo movimento sanitarista da Primeira

República. In *Raça, ciência e sociedade*, edited by Marcos Chor Maio and Ricardo Ventura Santos, 23–40. Rio de Janeiro: Editora Fiocruz.

Lima, Nísia Trindade, Ricardo Ventura Santos, and Carlos Everaldo Álvares Coimbra Jr. 2008. Rondonia de Edgard Roquette-Pinto: Antropologia e projeto nacional. In *Antropologia brasiliana: Ciência e educação na obra de Edgard Roquette-Pinto*, edited by Nísia Trindade Lima and Dominich Miranda de Sá, 99–121. Belo Horizonte, Brasil: Editora UFMG and Editora Fiocruz.

Lindee, Susan, and Ricardo Ventura Santos. 2012. The biological anthropology of living human populations: World histories, national styles and international networks. *Current Anthropology* 53(S5): S3–S16.

Lippman, Abby. 1991. Prenatal genetic testing and screening: Constructing needs and reinforcing inequities. *American Journal of Law and Medicine* 17(1–2): 15–50.

Lisker, Rubén. 1981. *Estructura genética de la población mexicana: Aspectos médicos y antropológicos*. Mexico City: Salvat Mexicana de Ediciones.

Lisker, Rubén, R. Perez-Briceño, J. Granados, et al. 1986. Gene frequencies and admixture estimates in a Mexico City population. *American Journal of Physical Anthropology* 71(2): 203–207.

Lisker, Rubén, E. Ramírez, and V. Babinsky. 1996. Genetic structure of autochthonous populations of Meso-America: Mexico. *Human Biology* 68(3): 395–404.

Lomnitz, Claudio. 2010a. *El antisemitismo y la ideología de la Revolución mexicana (Centzontle)*. Mexico City: Fondo de Cultura Económica.

Lomnitz, Claudio. 2010b. Los orígenes de nuestra supuesta homogeneidad: Breve arqueología de la unidad nacional en México. *Prismas* 14(1): 17–36.

Lomnitz, Claudio. 2010c. Por mi raza hablará el nacionalismo revolucionario (Arqueología de la unidad nacional). *Nexos*, 2 January, 42–51.

Lomnitz-Adler, Claudio. 1992. *Exits from the labyrinth: Culture and ideology in the Mexican national space*. Berkeley: University of California Press.

Lopes, Maria Margareth. 1993. *As ciências naturais e os museus no Brasil no século XIX*. PhD diss., Universidade de São Paulo.

López Beltrán, Carlos. 2004. *El sesgo hereditario: Ámbitos históricos del concepto de herencia biológica*. Mexico City: Universidad Nacional Autónoma de México.

López Beltrán, Carlos. 2007. Hippocratic bodies: Temperament and castas in Spanish America (1570–1820). *Journal of Spanish Cultural Studies* 8(2): 253–289.

López Beltrán, Carlos. 2008. Sangre y temperamento: Pureza y mestizajes en las sociedades de castas americanas. In *Saberes locales: Ensayos sobre historia de la ciencia en América Latina*, edited by Carlos López Beltrán and Frida Gorbach, 289–342. Zamora, Mexico: El Colegio de Michoacán.

López Beltrán, Carlos, ed. 2011. *Genes (y) mestizos: Genómica y raza en la biomedicina mexicana*. Mexico City: Ficticia Editorial.

López Beltrán, Carlos, and Francisco Vergara Silva. 2011. Genómica nacional: El INMEGEN y el genoma del mestizo. In *Genes (y) mestizos: Genómica y raza en la biomedicina mexicana*, edited by Carlos López Beltrán, 99–142. Mexico City: Ficticia Editorial.

López de Mesa, Luis. 1970 [1934]. *De cómo se ha formado la nación colombiana.* Medellín: Editorial Bedout.

Maca-Meyer, Nicole, Ana M. González, José M. Larruga, et al. 2001. Major genomic mitochondrial lineages delineate early human expansions. bmc *Genetics* 2(1).

Magnoli, Demétrio. 2009. *Uma gota de sangue: História do pensamento racial.* São Paulo: Editora Contexto.

Maio, Marcos Chor. 1999. Estoque semita: A presença dos judeus em *Casa Grande y Senzala. Luso-Brazilian Review* 36(1): 95–110.

Maio, Marcos Chor. 2001. U N E S C O and the study of race relations in Brazil: National or regional issue? *Latin American Research Review* 36: 118–136.

Maio, Marcos Chor, and Simone Monteiro. 2005. Tempos de racialização: O caso da "saúde da população negra" no Brasil. *História, Ciências, Saúde-Manguinhos* 12(2): 419–446.

Maio, Marcos Chor, and Simone Monteiro. 2010. Política social com recorte racial no Brasil: O caso da saúde da população negra. In *Raça como questão: História, ciência e identidades no Brasil,* edited by Marcos Chor Maio and Ricardo Ventura Santos, 285–314. Rio de Janeiro: Editora Fiocruz.

Maio, Marcos Chor, and Ricardo Ventura Santos, eds. 1996. *Raça, ciência e sociedade.* Rio de Janeiro: Editora Fiocruz.

Maio, Marcos Chor, and Ricardo Ventura Santos. 2005. Política de cotas raciais, os "olhos da sociedade" e os usos da antropologia: O caso do vestibular da Universidade de Brasília (UnB). *Horizontes Antropológicos* 11(23): 181–214.

Maio, Marcos Chor, and Ricardo Ventura Santos, eds. 2010. *Raça como questão: História, ciência e identidades no Brasil.* Rio de Janeiro: Editora Fiocruz.

Mallon, Florencia E. 1995. *Peasant and nation: The making of postcolonial Mexico and Peru.* Berkeley: University of California Press.

Mallon, Florencia E. 1996. Constructing *mestizaje* in Latin America: Authenticity, marginality and gender in the claiming of ethnic identities. *Journal of Latin American Anthropology* 2(1): 170–181.

Manolio, Teri A., Francis S. Collins, Nancy J. Cox, et al. 2009. Finding the missing heritability of complex diseases. *Nature* 461(7265): 747–753.

Mao, Xianyun, Abigail W. Bigham, Rui Mei, et al. 2007. A genomewide admixture mapping panel for Hispanic/Latino populations. *American Journal of Human Genetics* 80(6): 1171–1178.

Marks, Jonathan. 1995. *Human biodiversity: Genes, race, and history.* New Brunswick, NJ: Aldine Transaction.

Marks, Jonathan. 1996. The legacy of serological studies in American physical anthropology. *History and Philosophy of the Life Sciences* 18: 345–362.

Marks, Jonathan. 2001. "We're going to tell these people who they really are": Science and relatedness. In *Relative values: Reconfiguring kinship studies,* edited by Sarah Franklin and Susan McKinnon, 355–383. Durham, NC: Duke University Press.

Marrero, Andrea Rita. 2003. Os Gaúchos: Sua história evolutiva revelada a partir de marcadores genéticos. MA thesis, Universidade Federal do Rio Grande do Sul, Porto Alegre.

Marrero, Andrea Rita, Claudio Bravi, Steven Stuart, et al. 2007. Pre- and post-Columbian gene and cultural continuity: The case of the *Gaucho* from southern Brazil. *Human Heredity* 64(3): 160–71.

Marrero, Andrea Rita, Fábio Pereira Das Neves Leite, Bianca De Almeida Carvalho, et al. 2005. Heterogeneity of the genome ancestry of individuals classified as white in the state of Rio Grande do Sul, Brazil. *American Journal of Human Biology* 17(4): 496–506.

Marrero, Andrea Rita, Wilson A. Silva-Junior, Claudia M. Bravi, et al. 2007. Demographic and evolutionary trajectories of the Guarani and Kaingang natives of Brazil. *American Journal of Physical Anthropology* 132(2): 301–310.

Martínez, María Elena. 2008. *Genealogical fictions: Limpieza de sangre, religion, and gender in colonial Mexico.* Stanford, CA: Stanford University Press.

Martinez-Alier [Stolcke], Verena. 1989 [1974]. *Marriage, colour and class in nineteenth-century Cuba: A study of racial attitudes and sexual values in a slave society.* 2nd ed. Ann Arbor: University of Michigan Press.

Martínez-Cortés, G., I. Nuño-Arana, R. Rubi-Castellanos, et al. 2010. Origin and genetic differentiation of three Native Mexican groups (Purépechas, Triquis and Mayas): Contribution of CODIS-STRs to the history of human populations of Mesoamerica. *Annals of Human Biology* 37(6): 801–819.

Marx, Anthony. 1998. *Making race and nation: A comparison of South Africa, the United States, and Brazil.* Cambridge: Cambridge University Press.

Massin, Benoit. 1996. From Virchow to Fischer: Physical anthropology and "modern race theories" in Wilhelmine Germany. In *Volksgeist as method and ethic: Essays on Boasian ethnography and the German anthropological tradition*, edited by George W. Stocking Jr., 79–154. Madison: University of Wisconsin Press.

Matory, J. Lorand. 2006. The "New World" surrounds an ocean: Theorizing the live dialogue between African and African American cultures. In *Afro-Atlantic dialogues: Anthropology in the diaspora*, edited by Kevin Yelvington, 151–192. Santa Fe, NM: School of American Research Press.

Mayr, Ernst. 1982. *The growth of biological thought: Diversity, evolution, and inheritance.* Cambridge, MA: Harvard University Press.

Mazumdar, Pauline M. H. 1995. *Species and specificity: An interpretation of the history of immunology.* Cambridge: Cambridge University Press.

McClellan, Jon, and Mary-Claire King. 2010. Genetic heterogeneity in human disease. *Cell* 141(2): 210–217.

M'charek, Amade. 2000. Technologies of population: Forensic DNA testing practices and the making of differences and similarities. *Configurations* 8(1): 121–158.

M'charek, Amade. 2005a. *Human Genome Diversity Project: An ethnography of scientific practice.* Cambridge, MA: Cambridge University Press.

M'charek, Amade. 2005b. The mitochondrial Eve of modern genetics: Of peoples and genomes, or the routinization of race. *Science as Culture* 14(2): 161–183.

Medina, Andrés, and Carlos García Mora. 1983. *La quiebra política de la antropología social en México: La polarización (1971–1976)*. Mexico City: UNAM.

Mendoza, Roberto, Ignacio Zarante, and Gustavo Valbuena. 1997. *Aspectos demográficos de las poblaciones indígenas, negras y aisladas visitadas por la Gran Expedición Humana*. Terrenos de la Gran Expedición Humana. Serie de Reportes de Investigación 6. Bogotá: Pontificia Universidad Javeriana.

Mesa, Natalia R., María C. Mondragón, Iván D. Soto, et al. 2000. Autosomal, mtDNA, and Y-chromosome diversity in Amerinds: Pre- and post-Columbian patterns of gene flow in South America. *American Journal of Human Genetics* 67(5): 1277–1286.

Miller, Marilyn Grace. 2004. *Rise and fall of the cosmic race: The cult of mestizaje in Latin America*. Austin: University of Texas Press.

Mjoen, Jon Alfred. 1931. Cruzamento de raças. *Boletín de Eugenía* 3(32): 1–6.

Mol, Annemarie. 2003. *The body multiple: Ontology in medical practice*. Durham, NC: Duke University Press.

Montagu, Ashley. 1942. *Man's most dangerous myth: The fallacy of race*. New York: Columbia University Press.

Monteiro, John M. 1996. As "raças" indígenas no pensamento brasileiro do Império. In *Raça, ciência e sociedade*, edited by Ricardo Ventura Santos and Marcos Chor Maio, 15–22. Rio de Janeiro: Editora Fiocruz.

Montoya, Michael J. 2007. Bioethnic conscription: Genes, race, and Mexicana/o ethnicity in diabetes research. *Cultural Anthropology* 22(1): 94–128.

Montoya, Michael J. 2011. *Making the Mexican diabetic: Race, science, and the genetics of inequality*. Berkeley: University of California Press.

Morning, Ann. 2008. Ethnic classification in global perspective: A cross-national survey of the 2000 census round. *Population Research and Policy Review* 27(2): 239–272.

Mosquera Rosero-Labbé, Claudia, and Ruby Ester León Díaz, eds. 2010. *Acciones afirmativas y ciudadanía diferenciada étnico-racial negra, afrocolombiana, palenquera y raizal: Entre Bicentenarios de las Independencias y Constitución de 1991*. Bogotá: Universidad Nacional de Colombia.

Mothelet, Veronica Guerrero, and Stephan Herrera. 2005. Mexico launches bold genome project. *Nature Biotechnology* 23(9): 1030.

Moutinho, Laura. 2004. *Razão, "cor" e desejo: Uma análise comparativa sobre relacionamentos afetivo-sexuais "inter-raciais" no Brasil e África do Sul*. São Paulo: Editora da UNESP.

Munanga, Kabengele. 1999. *Rediscutindo a mestiçagem no Brasil: Identidade nacional versus identidade negra*. Petrópolis: Editora Vozes.

Nagel, Joane. 2003. *Race, ethnicity, and sexuality: Intimate intersections, forbidden frontiers*. Oxford: Oxford University Press.

Nascimento, Alexandre do, et al. 2008. 120 anos da luta pela igualdade racial no Brasil: Manifesto em defesa da justiça e constitucionalidade das cotas. Accessed 22 July 2013. http://media.folha.uol.com.br/cotidiano/2008/05/13/stf_manifesto_13_maio_2008.pdf.

Nash, Catherine. 2012a. Gendered geographies of genetic variation: Sex, gender and mobility in human population genetics. *Gender, Place and Culture* 19(4): 409–428.

Nash, Catherine. 2012b. Genetics, race and relatedness: Human mobility and difference in the Genographic Project. *Annals of the Association of American Geographers* 102: 1–18.

Nash, Catherine. 2012c. Genome geographies: Mapping national ancestry and diversity in human population genetics. *Transactions of the Institute of British Geographers* 38(2): 193–206.

Navarrete, Federico. 2004. *Las relaciones interétnicas en México*. Mexico City: Universidad Nacional Autónoma de México.

Nelkin, Dorothy, and M. Susan Lindee. 1996. *The dna mystique: The gene as a cultural icon*. New York: Freeman.

Nelson, Alondra. 2008a. Bio science: Genetic genealogy testing and the pursuit of African ancestry. *Social Studies of Science* 38(5): 759–783.

Nelson, Alondra. 2008b. The factness of diaspora: The social sources of genetic genealogy. In *Revisiting race in a genomic age*, edited by Barbara A. Koenig, Sandra Soo-Jin Lee, and Sarah S. Richardson, 253–268. New Brunswick, NJ: Rutgers University Press.

Nelson, Diane M. 1999. *A finger in the wound: Body politics in quincentennial Guatemala*. Berkeley: University of California Press.

Nelson, Matthew R., Daniel Wegmann, Margaret G. Ehm, et al. 2012. An abundance of rare functional variants in 202 drug target genes sequenced in 14,002 people. *Science* 337(6090): 100–104.

Oehmichen, Cristina. 2003. La multiculturalidad de la ciudad de México y los derechos indígenas. *Revista Mexicana de Ciencias Políticas y Sociales* (189): 147–169.

Olarte Sierra, María Fernanda. 2010. Achieving the desirable nation: Abortion and antenatal testing in Colombia: The case of amniocentesis. PhD diss., University of Amsterdam.

Olarte Sierra, María Fernanda, and Adriana Díaz del Castillo H. 2013. "We are all the same, we all are mestizo": On populations, nations, and discourses in genetics research in Colombia. *Science as Culture*, 18 October 2003.

Oliveira, Lúcia Lippi. 1990. *A questão nacional na primeira república*. São Paulo: Brasiliense.

Oliven, Ruben George. 2006. *A parte e o todo: A diversidade cultural no Brasil-nação*. 2nd ed. Coleção Identidade brasileira. Petrópolis: Editora Vozes.

Ordóñez Vásquez, Adriana, ed. 2000. *Variación biológica y cultural en Colombia*. Vol. 1, *Geografía humana de Colombia*. Bogotá: Instituto Colombiano de Cultura Hispánica.

Orlove, Benjamin. 1998. Down to earth: Race and substance in the Andes. *Bulletin of Latin American Research* 17(2): 207–222.

Ottensooser, Friedrich. 1944. Cálculo do grau de mistura racial através dos grupos sangüíneos. *Revista Brasileira de Biologia* 4: 531–537.

Ottensooser, Friedrich. 1962. Analysis of trihybrid populations. *American Journal of Human Genetics* 14: 278–280.

Palha, Teresinha de Jesus Brabo Ferreira, Elzemar Martins Ribeiro-Rodrigues, Ândrea Ribeiro-dos-Santos, et al. 2011. Male ancestry structure and interethnic admixture in African-descent communities from the Amazon as revealed by Y-chromosome STRs. *American Journal of Physical Anthropology* 144(3): 471–478.

Pallares-Burke, Maria Lúcia Garcia. 2005. *Gilberto Freyre: Um vitoriano dos tropicos.* São Paulo: Editora UNESP.

Palmié, Stephan. 2007. Genomics, divination, "racecraft." *American Ethnologist* 34(2): 205–222.

Pálsson, Gísli. 2007. *Anthropology and the new genetics.* Cambridge: Cambridge University Press.

Pálsson, Gísli. 2008. Genomic anthropology: Coming in from the cold? *Current Anthropology* 49(4): 545–568.

Pálsson, Gísli. 2009. Biosocial relations of production. *Comparative Studies in Society and History* 51(2): 288–313.

Paredes, Manuel, Aida Galindo, Margarita Bernal, et al. 2003. Analysis of the CODIS autosomal STR loci in four main Colombian regions. *Forensic Science International* 137(1): 67–73.

Parra, Flavia C., Roberto C. Amado, José R. Lambertucci, et al. 2003. Color and genomic ancestry in Brazilians. *Proceedings of the National Academy of Sciences* 100(1): 177–182.

Patterson, Nick, Alkes L. Price, and David Reich. 2006. Population structure and eigenanalysis. *PLoS Genetics* 2(12): e190.

Paz, Octavio. 1950. *El laberinto de la soledad.* Mexico City: Cuadernos Americanos.

Pena, S. D. J., L. Bastos-Rodrigues, J. R. Pimenta, et al. 2009. DNA tests probe the genomic ancestry of Brazilians. *Brazilian Journal of Medical and Biological Research* 42(10): 870–876.

Pena, Sérgio D. J., ed. 2002. *Homo brasilis: Aspectos genéticos, lingüísticos, históricos e socioantropológicos da formação do povo brasileiro.* Ribeirão Preto: FUNPEC-RP.

Pena, Sérgio D. J. 2005. Razões para banir o conceito de raça da medicina brasileira. *História, Ciências, Saúde-Manguinhos* 12(2): 321–346.

Pena, Sérgio D. J. 2007. Brazilians and the variable mosaic genome paradigm. In *Fifty years of human genetics: A festschrift and liber amicorum to celebrate the life and work of George Robert Fraser,* edited by Oliver Mayo and Carolyn Leach, 98–104. Adelaide: Wakefield.

Pena, Sérgio D. J. 2008. *Humanidade sem raças?* São Paulo: Publifolha.

Pena, Sérgio D. J., Denise R. Carvalho-Silva, Juliana Alves-Silva, et al. 2000. Retrato molecular do Brasil. *Ciência Hoje* 159: 16–25.

Pena, Sérgio D. J., and Maria Cátira Bortolini. 2004. Pode a genética definir quem deve se beneficiar das cotas universitárias e demais ações afirmativas? *Estudos Avançados* 18(50): 31–50.

Pena, Sérgio Danilo. 2006. Ciência, bruxas e raça. *Folha de São Paulo,* São Paulo, 2 August. Accessed 23 June 2010. http://www.jornaldaciencia.org.br/Detalhe .jsp?id=39579.

Pena, Sérgio Danilo. 2008. *Humanidade sem raças?* São Paulo: Publifolha (série 21).

Pena, Sérgio Danilo, Giuliano Di Pietro, Mateus Fuchshuber-Moraes, et al. 2011. The genomic ancestry of individuals from different geographical regions of Brazil is more uniform than expected. *Plos One* 6: e17063.

Penchaszadeh, Victor B. 2011. Forced disappearance and suppression of identity of children in Argentina: Experiences in genetic identification. In *Racial identities, genetic ancestry, and health in South America: Argentina, Brazil, Colombia, and Uruguay*, edited by Sahra Gibbon, Ricardo Ventura Santos, and Mónica Sans, 213–243. New York: Palgrave Macmillan.

Petryna, Adriana. 2009. *When experiments travel: Clinical trials and the global search for human subjects*. Princeton, NJ: Princeton University Press.

Pick, Daniel. 1989. *Faces of degeneration: A European disorder, c. 1848–1918*. Cambridge: Cambridge University Press.

Pimenta, Juliana R., Luciana W. Zuccherato, Adriana A. Debes, et al. 2006. Color and genomic ancestry in Brazilians: A study with forensic microsatellites. *Human Heredity* 62(4): 190–195.

Poole, Stafford. 1999. The politics of limpieza de sangre: Juan de Ovando and his circle in the reign of Philip II. *Americas* 55(3): 359–389.

Prado, Paulo. 1928. *Retrato do Brasil: Ensaio sobre a tristeza brasileira*. São Paulo: Oficinas Gráficas Duprat-Mayença.

Price, Alkes L., Nick Patterson, Fuli Yu, et al. 2007. A genomewide admixture map for Latino populations. *American Journal of Human Genetics* 80(6): 1024–1036.

Pritchard, J. K., M. Stephens, and P. Donnelly. 2000. Inference of population structure using multilocus genotype data. *Genetics* 155(2): 945–959.

Procesoscensales. 2011. Raza y racismo en Colombia (video, 3 parts). http://www.youtube.com/watch?v=LDHXls8wduo&p=292C776DB8B3121B.

Provine, William B. 1973. Geneticists and the biology of race crossing. *Science* 182: 790–796.

Provine, William B. 1986. Geneticists and race. *American Zoologist* 26: 857–887.

PUJ. 1992. A la zaga de la América oculta: Gran Expedición Humana 1992. *Innovación y Ciencia* 1(1): 14–19.

Rabinow, Paul. 1992. Artificiality and the Enlightenment: From sociobiology to biosociality. In *Incorporations*, edited by Jonathan Crary and Sanford Kwinter, 234–252. New York: Zone.

Rabinow, Paul. 1996. *Essays on the anthropology of reason*. Princeton, NJ: Princeton University Press.

Rabinow, Paul. 1999. *French dna: Trouble in purgatory*. Chicago: University of Chicago Press.

Rahier, Jean Muteba. 2003. Introduction: Mestizaje, mulataje, mestiçagem in Latin American ideologies of national identities. *Journal of Latin American Anthropology* 8(1): 40–50.

Rahier, Jean Muteba. 2011. From invisibilidad to participation in state corporatism: Afro-Ecuadorians and the constitutional processes of 1998 and 2008. *Identities: Global Studies in Power and Culture* 18(5): 502–527.

Ramos, Alcida. 1998. *Indigenism: Ethnic politics in Brazil*. Madison: University of Wisconsin Press.

Ramos, Catherine. 2004. Controversia en torno al proyecto "expedición humana" del Instituto de Genética Humana de la Universidad Javeriana: ¿Sangre para DracUSA? Undergraduate thesis, Universidad Nacional de Colombia, Bogotá.

Rappaport, Joanne. 2005. *Intercultural utopias: Public intellectuals, cultural experimentation and ethnic pluralism in Colombia*. Durham, NC: Duke University Press.

Rappaport, Joanne. 2012. Buena sangre y hábitos españoles: Repensando a Alonso de Silva y Diego de Torres. *Anuario Colombiano de Historia Social y de la Cultura* 39(1): 19–48.

Reardon, Jenny. 2001. The Human Genome Diversity Project: A case study in coproduction. *Social Studies of Science* 31(3): 357–388.

Reardon, Jenny. 2005. *Race to the finish: Identity and governance in an age of genomics*. Princeton, NJ: Princeton University Press.

Reardon, Jenny. 2008. Race without salvation: Beyond the science/society divide in genomic studies of human diversity. In *Revisiting race in a genomic age*, edited by Barbara A. Koenig, Sandra Soo-Jin Lee, and Sarah S. Richardson, 304–319. New Brunswick, NJ: Rutgers University Press.

Renique, Gerardo. 2003. Sonora's anti-Chinese racism and Mexico's postrevolutionary nationalism, 1920s–1930s. In *Race and nation in modern Latin America*, edited by Nancy P. Appelbaum, Anne S. Macpherson, and Karin A. Rosemblatt, 212–236. Chapel Hill: University of North Carolina Press.

Restrepo, Eduardo. 2007. Imágenes del "negro" y nociones de raza en Colombia a principios del siglo XX. *Revista de Estudios Sociales* 27: 46–61.

Restrepo, Eduardo. 2012. *Intervenciones en teoría cultural*. Popayán: Editorial Universidad del Cauca.

Restrepo, Eduardo, and Axel Rojas, eds. 2004. *Conflicto e (in)visibilidad: Retos de los estudios de la gente negra en Colombia*. Popayán, Colombia: Editorial Universidad del Cauca.

Restrepo, Eduardo, and Axel Rojas. 2010. *Inflexión decolonial: Fuentes, conceptos y cuestionamientos*. Popayán: Editorial Universidad del Cauca.

Rey, Mauricio, José Andrés Gutiérrez, Blanca Schroeder, et al. 2003. Allele frequencies for 13 STR's from two Colombian populations: Bogotá and Boyacá. *Forensic Science International* 136(1–3): 83–85.

Reynolds, Larry T., and Leonard Lieberman, eds. 1996. *Race and other misadventures: Essays in honor of Ashley Montagu in his ninetieth year*. Dix Hills, NY: General Hall.

Ribeiro, Guilherme Galvarros Bueno Lobo, Reginaldo Ramos De Lima, Cláudia Emília Vieira Wiezel, et al. 2009. Afro-derived Brazilian populations: Male genetic constitution estimated by Y-chromosomes STRs and AluYAP element polymorphisms. *American Journal of Human Biology* 21(3): 354–356.

Ribeiro, Silvia. 2005. El mapa genómico de los mexicanos. *La Jornada*, 31 July.

Ribeiro-Rodrigues, Elzemar, Teresinha de Palha, Eloisa Bittencourt, et al. 2011. Extensive survey of 12 X-STRs reveals genetic heterogeneity among Brazilian populations. *International Journal of Legal Medicine* 125(3): 445–452.

Roberts, Dorothy. 2010. Race and the new biocitizen. In *What's the use of race? Modern governance and the biology of difference*, edited by Ian Whitmarsh and David S. Jones, 259–276. Cambridge, MA: MIT Press.

Rodrigues, Elzemar Martins Ribeiro, Teresinha de Jesus Brabo Ferreira Palha, and Sidney Emanuel Batista dos Santos. 2007. Allele frequencies data and statistic parameters for 13 STR loci in a population of the Brazilian Amazon Region. *Forensic Science International* 168(2–3): 244–247.

Rodríguez Garavito, César, Tatiana Alfonso Sierra, and Isabel Cavelier Adarve. 2009. *Informe sobre discriminación racial y derechos de la población afrocolombiana: Raza y derechos humanos en Colombia*. Bogotá: Universidad de los Andes, Facultad de Derecho, Centro de Investigaciones Sociojurídicas, Observatorio de Discriminación Racial, Ediciones Uniandes.

Rojas, Madelyn, Angela Alonso, Leonardo Eljach, et al. 2012. Análisis de la estructura de la población de La Guajira: Una visión genética, demográfica y genealógica. In William Usaquén, Validación y consistencia de información en estudios de diversidad genética humana a partir de marcadores microsatélites (PhD diss.), 63–101. Bogotá: Universidad Nacional de Colombia.

Rojas, Winston, María Victoria Parra, Omer Campo, et al. 2010. Genetic make up and structure of Colombian populations by means of uniparental and biparental DNA markers. *American Journal of Physical Anthropology* 143(1): 13–20.

Romero, Rosa Elena, Ignacio Briceño, Rocío del Pilar Lizarazo, et al. 2008. A Colombian Caribbean population study of 16 Y-chromosome STR loci. *Forensic Science International: Genetics* 2(2): e5–e8.

Rondón, Fernando, Julio César Osorio, Ángela Viviana Peña, et al. 2008. Diversidad genética en poblaciones humanas de dos regiones colombianas. *Colombia Médica* 39(2, Suppl. 2): 52–60.

Roquette-Pinto, Edgard. 1917. *Rondonia (anthropologia; ethnographia)*. Archivos do Museu Nacional 20. Rio de Janeiro: Imprensa Nacional.

Roquette-Pinto, Edgard. 1927. *Seixos rolados*. Rio de Janeiro: Mendonça, Machado.

Roquette-Pinto, Edgard. 1929. Nota sobre os typos anthropologicos do Brasil. In *Actas e trabalhos do primeiro congresso brasileiro de eugenia*, 1, Rio de Janeiro.

Roquette-Pinto, Edgard. 1933. *Ensaios de anthropologia brasiliana*. São Paulo: Companhia Editora Nacional.

Roquette-Pinto, Edgard. 1942. *Ensaios brasilianos*. São Paulo: Cia Editora Nacional.

Rose, Nikolas. 2007. *The politics of life itself: Biomedicine, power and subjectivity in the twenty-first century*. Princeton, NJ: Princeton University Press.

Rose, Nikolas, and Carlos Novas. 2005. Biological citizenship. In *Global assemblages: Technology, politics, and ethics as anthropological problems*, edited by Aihwa Ong and Stephen J. Collier, 439–463. Oxford: Blackwell.

Ruiz-Linares, Andrés, Daniel Ortíz-Barrientos, Mauricio Figueroa, et al. 1999. Microsatellites provide evidence for Y chromosome diversity among the founders of the New World. *Proceedings of the National Academy of Sciences* 96(11): 6312–6317.

Ruppert, Evelyn. 2009. Numbers regimes: From censuses to metrics. CRESC *Working Papers Series* 68. Milton Keynes, U.K.: Open University.

Rusnock, Andrea. 1997. Quantification, precision, and accuracy: Determinations of population in the Ancien Régime. In *The values of precision*, edited by M. Norton Wise, 17–38. Princeton, NJ: Princeton University Press.

Rutsch, Mechthild. 2007. *Entre el campo y el gabinete: Nacionales y extranjeros en la profesionalización de la antropología mexicana (1877–1920)*. Mexico City: Instituto Nacional de Antropología e Historia/Universidad Nacional Autónoma de México.

Sá, Guilherme José da Silva e, Ricardo Ventura Santos, Claudia Rodrigues-Carvalho, et al. 2008. Crânios, corpos e medidas: A constituição do acervo de instrumentos antropométricos do Museu Nacional na passagem do século XIX para o XX. *História, Ciências, Saúde-Manguinhos* 15(1): 197–208.

Saade Granados, Marta. 2009. El mestizo no es "de color." Ciencia y política pública mestizófilas (México, 1920–1940). PhD diss., Escuela Nacional de Antropología e Historia ENAH–INAH, Mexico City.

Safier, Neil. 2010. Global knowledge on the move: Itineraries, Amerindian narratives, and deep histories of science. *Isis* 101(1): 133–145.

Sahlins, Marshall. 1999. Two or three things that I know about culture. *Journal of the Royal Anthropological Institute* 5(3): 399–421.

Sala de Prensa del Gobierno Federal. 2009. Discurso del Presidente Calderón en la presentación del Mapa del Genoma de los Mexicanos. Unpublished press release, 11 May.

Salas, Alberto. 1960. *Crónica florida del mestizaje de las Indias siglo XVI*. Buenos Aires: Ed. Losada.

Salek, Silvia. 2007. BBC delves into Brazilians' roots. BBC World News, 10 July. http://news.bbc.co.uk/1/hi/6284806.stm.

Salzano, Francisco M., ed. 1971. *The ongoing evolution of Latin American populations*. Springfield, IL: Thomas.

Salzano, Francisco M. 1997. Human races: Myth, invention, or reality? *Interciencia* 22(5): 221–227.

Salzano, Francisco M., ed. 2011. *Recordar é viver: A história da Sociedade Brasileira de Genética*. Ribeirão Preto: Sociedade Brasileira de Genética.

Salzano, Francisco M., and Sídia Maria Callegari-Jacques. 1988. *South American Indians: A case study in evolution*. Oxford: Clarendon.

Salzano, Francisco M., and Maria Cátira Bortolini. 2002. *The evolution and genetics of Latin American populations*. Cambridge: Cambridge University Press.

Salzano, Francisco M., and Maria Cátira Bortolini. 2003. Genes dos gaúchos para deduzir a história genética da América e a evolução de sua ocupação nativa. Canal Ciencia. Accessed 29 July 2013. http://www.canalciencia.ibict.br/pesquisa/0162-Genes-dos-gauchos-e-a-historia-genetica-das-americas.html.

Salzano, Francisco M., and Newton Freire-Maia. 1967. *Populações brasileiras: Aspectos demográficos, genéticos e antropológicos*. São Paulo: Editora Nacional/EDUSP.

Salzano, Francisco M., and Newton Freire-Maia. 1970. *Problems in human biology: A study of Brazilian populations*. Detroit: Wayne State University Press.

Samper, José María. 1861. *Ensayo sobre las revoluciones políticas y la condición social de* las repúblicas colombianas (hispano-americanas): Con un apéndice sobre la orografía y la población de la Confederación Granadina. Paris: Imprenta de E. Thunot y Cia.

Sánchez-Giron, Francisco, Beatriz Villegas-Torres, Karla Jaramillo-Villafuerte, et al. 2011. Association of the genetic marker for abacavir hypersensitivity HLA-B*5701 with HCP5 rs2395029 in Mexican Mestizos. Pharmacogenomics 12(6): 809–814.

Sánchez-Guillermo, Evelyne. 2007. Nacionalismo y racismo en el México decimonónico: Nuevos enfoques, nuevos resultados. Nuevo Mundo Mundos Nuevos. Accessed 13 September 2012. http://nuevomundo.revues.org/3528.

Sanders, James. 2004. Contentious republicans: Popular politics, race, and class in nineteenth-century Colombia. Durham, NC: Duke University Press.

Sandoval, C., A. De la Hoz, and E. Yunis. 1993. Estructura genética de la población colombiana: Análisis de mestizaje. Revista Facultad de Medicina de la Universidad Nacional de Colombia 41: 3–14.

Sanín, Javier. 1992. Editorial: La Gran Expedición Humana. Revista Javeriana 118(586): 7–8.

Sans, Monica. 2000. Admixture studies in Latin America: From the 20th to the 21st century. Human Biology 72(1): 155–177.

Sans, Mónica, Gonzalo Figueiro, Carlos Sanguinetti, et al. 2010. The "last Charrúa Indian" (Uruguay): Analysis of the remains of Chief Vaimaca Perú. Nature Precedings. Accessed 25 March 2013. http://precedings.nature.com/documents/4415/version/1/files/npre20104415-1.pdf.

Sansone, Lívio. 2003. Blackness without ethnicity: Constructing race in Brazil. New York: Palgrave Macmillan.

Santos, Eduardo José Melo dos, Andrea Kelly Campos Ribeiro-dos-Santos, João Farias Guerreiro, et al. 1996. Migration and ethnic change in an admixed population from the Amazon region (Santarém, Pará). Brazilian Journal of Genetics 19(3): 511–515.

Santos, Ney P. C., Elzemar M. Ribeiro-Rodrigues, Ândrea K. C. Ribeiro-dos-Santos, et al. 2009. Assessing individual interethnic admixture and population substructure using a 48-insertion-deletion (INSEL) ancestry-informative marker (AIM) panel. Human Mutation 31(2): 184–190.

Santos, Ricardo Ventura. 1998. A obra de Euclides da Cunha e os debates sobre mestiçagem no Brasil no início do século XX: Os Sertões e a medicina-antropologia do Museu Nacional. História, Ciência, Saúde-Manguinhos 5(Suppl.): 237–253.

Santos, Ricardo Ventura. 2002. Indigenous peoples, postcolonial contexts and genomic research in the late 20th century: A view from Amazonia (1960–2000). Critique of Anthropology 22(1): 81–104.

Santos, Ricardo Ventura. 2006. Indigenous peoples, bioanthropological research, and ethics in Brazil: Issues in participation and consent. In The nature of difference: Science, society and human biology, edited by George T. H. Ellison and Alan H. Goodman, 181–202. Boca Raton, FL: CRC Press.

Santos, Ricardo Ventura. 2012. Guardian angel on a nation's path: Contexts and trajectories of physical anthropology in Brazil in the late nineteenth and early twentieth centuries. *Current Anthropology* 53(s5): s17–s32.

Santos, Ricardo Ventura, and Marcos Chor Maio. 2004. Race, genomics, identity and politics in contemporary Brazil. *Critique of Anthropology* 24: 347–378.

Santos, Ricardo Ventura, and Marcos Chor Maio. 2005. Anthropology, race, and the dilemmas of identity in the age of genomics. *História, Ciências, Saúde-Manguinhos* 12: 447–468.

Santos, Ricardo Ventura, Peter H. Fry, Simone Monteiro, et al. 2009. Color, race, and genomic ancestry in Brazil: Dialogues between anthropology and genetics. *Current Anthropology* 50: 787–819.

Santos, S. E. B., J. D. Rodrigues, A. K. Ribeiro-dos-Santos, et al. 1999. Differential contribution of indigenous men and women to the formation of an urban population in the Amazon region as revealed by mtDNA and Y-DNA. *American Journal of Physical Anthropology* 109(2): 175–80.

Santos, Sidney Emanuel Batista dos, Ândrea Kely Campos Ribeiro dos Santos, Eduardo José Melo dos Santos, et al. 1999. The Amazon microcosm. *Ciência e Cultura* 51(3–4): 181–190.

Schneider, H., João Farias Guerreiro, Sidney Emanuel Batista dos Santos, et al. 1987. Isolate breakdown in Amazonia: The blacks of the Trombetas river. *Revista Brasileira de Genética* 10(3): 565–574.

Schramm, Katharina, David Skinner, and Richard Rottenburg, eds. 2011. *Identity politics and the new genetics: Re/creating categories of difference and belonging.* Oxford: Berghahn.

Schwaller, Robert C. 2011. "Mulata, hija de negro y india": Afro-indigenous mulatos in early colonial Mexico. *Journal of Social History* 44(3): 889–914.

Schwarcz, Lilia Moritz. 1993. *O espetáculo das raças: Cientistas, instituições e questão racial no Brasil, 1870–1930.* São Paulo: Companhia das Letras.

Schwartz, Ernesto. 2008. Genomic sovereignty and the creation of the INMEGEN: Governance, populations and territoriality. MA thesis, University of Exeter.

Schwartz-Marín, Ernesto. 2011. Genomic sovereignty and the "Mexican genome": An ethnography of postcolonial biopolitics. PhD diss., University of Exeter.

Schwartz-Marín, Ernesto, and Irma Silva-Zolezzi. 2010. "The map of the Mexican's genome": Overlapping national identity, and population genomics. *Identity in the Information Society* 3(3): 489–514.

Schwartz-Marín, Ernesto, Peter Wade, Eduardo Restrepo, Areli Cruz-Santiago, and Roosbelinda Cárdenas. 2013. Colombian forensic genetics as a form of public science: The role of race, nation and common sense in the stabilisation of DNA populations. Unpublished manuscript.

Seigel, Micol. 2009. *Uneven encounters: Making race and nation in Brazil and the United States.* Durham, NC: Duke University Press.

Serre, David, and Svante Pääbo. 2004. Evidence for gradients of human genetic diversity within and among continents. *Genome Research* 14(9): 1679–1685.

Seyferth, Giralda. 1985. A antropologia e a teoria do branqueamento da raça no Brasil: A tese de João Batista de Lacerda. *Revista do Museu Paulista* 30: 81–98.

Seyferth, Giralda. 2008. Roquette-Pinto e o debate sobre raça e imigração no Brasil. In *Antropologia brasiliana: Ciência e educação na obra de Edgard Roquette-Pinto*, edited by Nísia Trindade Lima and Dominichi Miranda de Sá, 147–177. Rio de Janeiro: Editora UFMG and Editora Fiocruz.

Shriver, Mark D., and Rick A. Kittles. 2008. Genetic ancestry and the search for personalized genetic histories. In *Revisiting race in a genomic age*, edited by Barbara A. Koenig, Sandra Soo-Jin Lee, and Sarah S. Richardson, 201–214. New Brunswick, NJ: Rutgers University Press.

Sicroff, Albert A. 1985. *Los estatutos de limpieza de sangre: Controversias entre los siglos XV y XVII*. Translated by Mauro Armiño. Madrid: Taurus.

Sieder, Rachel, ed. 2002. *Multiculturalism in Latin America: Indigenous rights, diversity and democracy*. Houndmills, U.K.: Palgrave Macmillan.

Silva-Zolezzi, Irma, Alfredo Hidalgo-Miranda, Jesus Estrada-Gil, et al. 2009. Analysis of genomic diversity in Mexican Mestizo populations to develop genomic medicine in Mexico. *Proceedings of the National Academy of Sciences* 106(21): 8611–8616.

Simpson, Bob. 2000. Imagined genetic communities: Ethnicity and essentialism in the twenty-first century. *Anthropology Today* 16(3): 3–6.

Sivasundaram, Sujit. 2010. Sciences and the global: On methods, questions, and theory. *isis* 101(1): 146–158.

Skidmore, Thomas. 1974. *Black into white: Race and nationality in Brazilian thought*. New York: Oxford University Press.

Skinner, David. 2006. Racialised futures: Biologism and the changing politics of identity. *Social Studies of Science* 36(3): 459–488.

Skinner, David. 2007. Groundhog day? The strange case of sociology, race and "science." *Sociology* 41(5): 931–943.

Skipper, Magdalena, Ritu Dhand, and Philip Campbell. 2012. Presenting ENCODE. *Nature* 489(7414): 45.

Smedley, Audrey. 1993. *Race in North America: Origin and evolution of a worldview*. Boulder: Westview.

Smith, Carol A. 1997. The symbolics of blood: Mestizaje in the Americas. *Identities: Global Studies in Power and Culture* 3(4): 495–521.

Souza, Vanderlei S., Carlos E. A. Coimbra Jr., Ricardo Ventura Santos, and Rodrigo C. Dornelles. 2013. História da genética no Brasil: Um olhar a partir do "Museu da Genética" da Universidade Federal do Rio Grande do Sul (UFRGS). *História, Ciências, Saúde-Manguinhos* 20(2): 675–694.

Souza-Lima, Antonio Carlos de. 1995. *Um grande cerco de paz: Poder tutelar, indianidade e formação do Estado no Brasil*. Petrópolis: Vozes.

Speed, Shannon. 2005. Dangerous discourses: Human rights and multiculturalism in neoliberal Mexico. *PoLAR: Political and Legal Anthropology Review* 28(1): 29–51.

Spencer, Frank, ed. 1997. *History of physical anthropology: An encyclopedia*. New York: Garland.

Star, Susan Leigh, and James R. Griesemer. 1989. Institutional ecology, "translations" and boundary objects: Amateurs and professionals in Berkeley's Museum of Vertebrate Zoology, 1907–39. *Social Studies of Science* 19(3): 387–420.

Star, Susan Leigh. 1991. Power, technology and the phenomenology of conventions: On being allergic to onions. In *A sociology of monsters: Essays on power, technology and dominations*, edited by John Law, 26–56. London: Routledge.

Steil, Carlos Alberto, ed. 2006. *Cotas raciais na universidade: Um debate.* Porto Alegre: Editora UFRGS.

Stengers, Isabelle. 2007. Diderot's egg: Divorcing materialism from eliminativism. *Radical Philosophy* 144(7–15).

Stepan, Nancy. 1982. *The idea of race in science: Great Britain, 1800–1960.* London: Macmillan.

Stepan, Nancy L. 1985. Biology and degeneration: Races and proper places. In *Degeneration: The dark side of progress*, edited by J. Edward Chamberlin and Sander L. Gilman, 97–120. New York: Columbia University Press.

Stepan, Nancy L. 1991. *"The hour of eugenics": Race, gender and nation in Latin America.* Ithaca, NY: Cornell University Press.

Stephens, Sharon. 1995. Physical and cultural reproduction in a post-Chernobyl Norwegian Sami community. In *Conceiving the new world order: The global politics of reproduction*, edited by Faye D. Ginsburg and Rayna Rapp, 270–288. Berkeley: University of California Press.

Stern, Alexandra. 2000. *Mestizophilia, biotypology, and eugenics in post-revolutionary Mexico: Towards a history of science and the state, 1920–1960.* Chicago: University of Chicago, Mexican Studies Program, Center for Latin American Studies.

Stern, Alexandra Minna. 2009. Eugenics and racial classification in modern Mexican America. In *Race and classification: The case of Mexican America*, edited by Ilona Katzew and Susan Deans-Smith, 151–173. Stanford, CA: Stanford University Press.

Stocking, George. 1982. *Race, culture and evolution: Essays on the history of anthropology.* 2nd ed. Chicago: University of Chicago Press.

Stocking, George W., Jr. 1968. *Race, culture and evolution: Essays in the history of anthropology.* New York: Free Press.

Stocking, George W., Jr., ed. 1988. *Bones, bodies, behavior: Essays on biological anthropology.* Madison: University of Wisconsin Press.

Stolcke, Verena. 1994. Invaded women: Gender, race, and class in the formation of colonial society. In *Women, "race," and writing in the early modern period*, edited by Margo Hendricks and Patricia Parker, 272–286. London: Routledge.

Stolcke, Verena. 1995. Talking culture: New boundaries, new rhetorics of exclusion in Europe. *Current Anthropology* 36(1): 1–23.

Strathern, Marilyn. 1992. *After nature: English kinship in the late twentieth century.* Cambridge: Cambridge University Press.

Stutzman, Ronald. 1981. El mestizaje: An all-inclusive ideology of exclusion. In *Cultural transformations and ethnicity in modern Ecuador*, edited by Norman E. Whitten, 45–94. Urbana: University of Illinois Press.

Suárez, Edna, and Ana Barahona. 2011. La nueva ciencia de la nación mestiza: Sangre y genética humana en la posrevolución mexicana (1945–1967). In *Genes (y) mestizos. Genómica y raza en la biomedicina mexicana*, edited by Carlos López Beltrán, 65–96. Mexico City: Ficticia Editorial.

Suarez-Kurtz, Guilherme. 2011. Pharmacogenetics in the Brazilian population. In *Racial identities, genetic ancestry, and health in South America: Argentina, Brazil, Colombia, and Uruguay*, edited by Sahra Gibbon, Ricardo Ventura Santos, and Mónica Sans, 121–135. New York: Palgrave Macmillan.

Suárez y López-Guaso, Laura, and Rosaura Ruíz-Gutiérrez. 2001. Eugenesia y medicina social en el México posrevolucionario. *Ciencias* 60–61: 80–97.

TallBear, Kim. 2007. Narratives of race and indigeneity in the Genographic Project. *Journal of Law, Medicine and Ethics* 35(3): 412–424.

Tandon, Arti, Nick Patterson, and David Reich. 2011. Ancestry informative marker panels for African Americans based on subsets of commercially available SNP arrays. *Genetic Epidemiology* 35(1): 80–83.

Tarica, Estelle. 2008. *The inner life of mestizo nationalism*. Minneapolis: University of Minnesota Press.

Taussig, Karen-Sue. 2009. *Ordinary genomes: Science, citizenship, and genetic identities*. Durham, NC: Duke University Press.

Taussig, Karen-Sue, Rayna Rapp, and Deborah Heath. 2003. Flexible eugenics: Technologies of the self in the age of genetics. In *Genetic nature/culture: Anthropology and science beyond the two-culture divide*, edited by Alan H. Goodman, Deborah Heath, and Susan M. Lindee, 58–76. Berkeley: University of California Press.

Telles, Edward E. 2003. US foundations and racial reasoning in Brazil. *Theory, Culture and Society* 20(4): 31–47.

Telles, Edward E. 2004. *Race in another America: The significance of skin color in Brazil*. Princeton, NJ: Princeton University Press.

Tennessen, Jacob A., Abigail W. Bigham, Timothy D. O'Connor, et al. 2012. Evolution and functional impact of rare coding variation from deep sequencing of human exomes. *Science* 337(6090): 64–69.

Tenorio Trillo, Mauricio. 2006. Del mestizaje a contrapelo: Guatemala y México. *Istor* 24: 67–94.

Teresa de Mier, Fray Servando. 1987 [1811]. Sobre el origen de los españoles y la mezcla de su sangre. In *Cartas de un americano al español, 1811–1813*. Mexico City: Secretaría de Educación Pública.

Terreros, Grace Alexandra. 2010. Determinación de la variación de las secuencias de las regiones HVI y HVII de la región control del DNA mitocondrial en una muestra de la población Caribe colombiana. MA thesis, Universidad Javeriana, Bogotá.

Thielen, Eduardo Vilela, Fernando A. P. Alves, and Jaime L. Benchimol. 1991. *A ciência a caminho da roça: Imagens das expedições científicas do Instituto Oswaldo Cruz (1903–1911)*. Rio de Janeiro: Editora Fiocruz.

Thomson, Sinclair. 2011. Was there race in colonial Latin America? Identifying selves and others in the insurgent Andes. In *Histories of race and racism: The Andes and Mesoamerica from colonial times to the present*, edited by Laura Gotkowitz, 72–91. Durham, NC: Duke University Press.

Torres Carvajal, María Mercedes. 2005. La variabilidad genética: Una herramienta útil en el estudio de poblaciones humanas. PhD diss., Universidad de los Andes, Bogotá.

Torroni, Antonio, James V. Neel, Ramiro Barrantes, et al. 1994. Mitochondrial DNA "clock" for the Amerinds and its implications for timing their entry into North America. *Proceedings of the National Academy of Sciences* 91(3): 1158–1162.

Tsing, Anna L. 1993. *In the realm of the diamond queen: Marginality in an out-of-the-way place*. Princeton: Princeton University Press.

Tutton, Richard. 2004. Person, property, and gift: Exploring languages of tissue donation to biomedical research. In *Genetic databases: Socio-ethical issues in the collection and use of dna*, edited by Oonagh Corrigan and Richard Tutton, 19–38. London: Routledge.

Twine, France W. 1998. *Racism in a racial democracy: The maintenance of white supremacy in Brazil*. New Brunswick, NJ: Rutgers University Press.

Uribe, Consuelo. 2010. Estudio sobre la interdisciplinariedad en la Universidad Javeriana. El caso de la Facultad de Estudios Interdisciplinarios. Unpublished manuscript.

Uribe, María Victoria, and Eduardo Restrepo, eds. 1997. *Antropología en la modernidad: Identidades, etnicidades y movimientos sociales en Colombia*. Bogotá: Instituto Colombiano de Antropología.

Usaquén, William. 2012. Validación y consistencia de información en estudios de diversidad genética humana a partir de marcadores microsatélites. PhD diss., Universidad Nacional de Colombia, Bogotá.

Van Cott, Donna Lee. 2000. *The friendly liquidation of the past: The politics of diversity in Latin America*. Pittsburgh: University of Pittsburgh Press.

Vargas, A. E., A. R. Marrero, F. M. Salzano, et al. 2006. Frequency of CCR5delta32 in Brazilian populations. *Brazilian Journal of Medical and Biological Research* 39(3): 321–325.

Vasconcelos, José. 1997 [1925]. *The cosmic race: A bilingual edition*. Translated by Didier T. Jaén. Baltimore, MD: Johns Hopkins University Press.

Vergara Silva, Francisco. 2013. "Un asunto de sangre": Juan Comas, el evolucionismo bio-info-molecularizado, y las nuevas vidas de la ideología indigenista en México. In *Miradas plurales al fenómeno humano*, edited by Josefina Mansilla Lory and Xavier Lizarrga Cruchaca. Mexico City: Instituto Nacional de Antropología e Historia.

Villalobos-Comparán, Marisela, M. Teresa Flores-Dorantes, M. Teresa Villarreal-Molina, et al. 2008. The FTO gene is associated with adulthood obesity in the Mexican population. *Obesity* 16(10): 2296–2301.

Villella, Peter B. 2011. "Pure and noble Indians, untainted by inferior idolatrous races": Native elites and the discourse of blood purity in late colonial Mexico. *Hispanic American Historical Review* 91(4): 633–663.

Villoro, Luis. 1950. *Los grandes momentos del indigenismo en México*. Mexico City: El Colegio de México.

Viqueira, Juan Pedro. 2010. Reflexiones contra la noción histórica de mestizaje. *Nexos*, 1 May, 76–83.

Wade, Peter. 1993. *Blackness and race mixture: The dynamics of racial identity in Colombia.* Baltimore, MD: Johns Hopkins University Press.

Wade, Peter. 1999. Representations of blackness in Colombian popular music. In *Representations of blackness and the performance of identities*, edited by Jean M. Rahier, 173–191. Westport, CT: Greenwood.

Wade, Peter. 2002a. The Colombian Pacific in perspective. *Journal of Latin American Anthropology* 7(2): 2–33.

Wade, Peter. 2002b. *Race, nature and culture: An anthropological perspective.* London: Pluto.

Wade, Peter. 2004. Images of Latin American mestizaje and the politics of comparison. *Bulletin of Latin American Research* 23(1): 355–366.

Wade, Peter. 2005. Rethinking mestizaje: Ideology and lived experience. *Journal of Latin American Studies* 37: 1–19.

Wade, Peter, ed. 2007a. *Race, ethnicity and nation: Perspectives from kinship and genetics.* Oxford: Berghahn.

Wade, Peter. 2007b. Race, ethnicity and nation: Perspectives from kinship and genetics. In *Race, ethnicity and nation: Perspectives from kinship and genetics*, edited by Peter Wade, 1–31. Oxford: Berghahn.

Wade, Peter. 2009. *Race and sex in Latin America.* London: Pluto.

Wade, Peter. 2010. *Race and ethnicity in Latin America.* 2nd ed. London: Pluto.

Wailoo, Keith, Alondra Nelson, and Catherine Lee, eds. 2012. *Genetics and the unsettled past: The collision of dna, race, and history.* New Brunswick, NJ: Rutgers University Press.

Wang, Sijia, Cecil M. Lewis Jr., Mattias Jakobsson, et al. 2007. Genetic variation and population structure in Native Americans. *PLoS Genetics* 3(11): e185.

Wang, Sijia, Nicolas Ray, Winston Rojas, et al. 2008. Geographic patterns of genome admixture in Latin American mestizos. *PLoS Genetics* 4(3): e1000037.

Warman, Arturo. 1970. *De eso que llaman antropología mexicana.* Mexico City: Editorial Nuestro Tiempo.

Weismantel, Mary. 2001. *Cholas and pishtacos: Stories of race and sex in the Andes.* Chicago: University of Chicago Press.

Whitmarsh, Ian, and David S. Jones, eds. 2010. *What's the use of race? Modern governance and the biology of difference.* Cambridge, MA: MIT Press.

Wimmer, Andreas, and Nina Glick Schiller. 2002. Methodological nationalism and beyond: Nation-state building, migration and the social sciences. *Global Networks: A Journal of Transnational Affairs* 2: 301–334.

Wise, M. Norton. 1997a. Introduction. In *The values of precision*, edited by M. Norton Wise, 3–13. Princeton, NJ: Princeton University Press.

Wise, M. Norton. 1997b. Precision: Agent of unity, product of agreement. In *The values of precision*, edited by M. Norton Wise, 352–361. Princeton, NJ: Princeton University Press.

Wise, M. Norton, ed. 1997c. *The values of precision*. Princeton, NJ: Princeton University Press.

Yang, Jian, Beben Benyamin, Brian P. McEvoy, et al. 2010. Common SNPs explain a large proportion of the heritability for human height. *Nature Genetics* 42(7): 565–569.

Yashar, Deborah. 2005. *Contesting citizenship in Latin America: The rise of indigenous movements and the postliberal challenge*. Cambridge: Cambridge University Press.

Yunis Turbay, Emilio. 2006. *¡Somos así!* Bogotá: Editorial Bruna.

Yunis Turbay, Emilio. 2009. *¿Por qué somos así? ¿Qué pasó en Colombia? Análisis del mestizaje*. 2nd ed. Bogotá: Temis.

Yunis, J. J., O. García, I. Uriarte, et al. 2000. Population data on 6 short tandem repeat loci in a sample of Caucasian-Mestizos from Colombia. *International Journal of Legal Medicine* 113(3): 175–178.

Yunis, Juan J., Luis E. Acevedo, David S. Campo, et al. 2005. Population data of Y-STR minimal haplotypes in a sample of Caucasian-Mestizo and African descent individuals of Colombia. *Forensic Science International* 151(2–3): 307–313.

Zarante, Ignacio. 2013. Cifras de la Gran Expedición Humana. Accessed 5 April 2013. www.javeriana.edu.co/Humana/cifras.html.

Zea, Leopoldo. 2002. La frontera en la globalización. In *Frontera y globalización*, edited by Leopoldo Zea and Hernán Taboada, 5–13. Mexico City: Tierra Firme/Fondo de Cultura Económica.

Zembrzuski, V. M., S. M. Callegari-Jacques, and M. H. Hutz. 2006. Application of an African ancestry index as a genomic control approach in a Brazilian population. *Annals of Human Genetics* 70: 822–828.

Zuñiga, Jean-Paul. 1999. La voix du sang: Du métis à l'idée de métissage en Amérique espagnole. *Annales: Histoire, Sciences Sociales* 54(2): 425–452.

contributors

Roosbelinda Cárdenas did her PhD in cultural anthropology at the University of California, Santa Cruz. She also has an MA in Latin American studies from the University of Texas at Austin. She is currently a postdoctoral associate at Rutgers University's Center for Race and Ethnicity. Her dissertation studied the articulations of blackness in the moment after Colombia's multicultural turn and the intensification of the armed conflict, which disproportionately affected black communities in the Pacific coastal region.

Adriana Díaz del Castillo H. holds an MA in medical anthropology from the University of Amsterdam and an MD from the Universidad Nacional de Colombia. She has worked with public and private institutions both in Colombia and abroad. Her research interests have to do with the interplay of health and society, using an ethnographic approach and working on diverse topics, including the body, chronic illness, urban infrastructures, and well-being. Recently she has worked in the field of social studies of science and technology and has participated in projects regarding human population genetics, forensics, and information systems, and their relation to nation-building processes in Colombia. She currently works as a consultant and a researcher with two research groups at the Universidad de los Andes in Bogotá.

Vivette García Deister did a BSc in biology at the Universidad Nacional Autónoma de México, followed by an MA in the philosophy of science, completed in 2005. In 2009, she finished her doctorate in philosophical and social studies of science and technology at UNAM. She is also an adjunct lecturer in history and philosophy of biology at UNAM and has spent periods as a research fellow at the Max Planck Institute for the History of Science, Berlin, and at the Philosophy Department, University of California–Davis. From 2010 to 2013 she was postdoctoral research fellow in Social Anthropology at the University of Manchester. She is currently associate professor in S&TS at UNAM's School of Sciences.

Verlan Valle Gaspar Neto has a doctorate in anthropology from the Universidade Federal Fluminense, with a dissertation "Biological Anthropology in Brazil Today" (2012). He has a master's in anthropology from the same university (2008), and a degree in social sciences from the Universidade Federal de Juiz de Fora

(2005). He is currently an assistant professor at the Federal University of Alfenas, in Minas Gerais state, Brazil.

Michael Kent is a social anthropologist, with an MA from the VU University Amsterdam and a PhD from the University of Manchester. His doctoral research focused on transformations in the relationship between indigenous movements and the state in the Peruvian Andes. Michael's current research explores the relations between genetic ancestry studies, public debates on social identity, and political conflict in Brazil, Peru, and Uruguay. He has worked as lecturer at the University of Amsterdam and the VU University Amsterdam. He is currently affiliated with Social Anthropology at the University of Manchester.

Carlos López Beltrán is a historian of science and senior researcher in the Instituto de Investigaciones Filosóficas, Universidad Nacional Autónoma de México. He has written extensively on the history of science in Mexico and on theories of heredity. His books include El sesgo hereditario, ámbitos históricos del concepto de herencia biológica.

María Fernanda Olarte Sierra has a PhD in social sciences, a master's degree in medical anthropology, and a BA in social anthropology. Her research interests have revolved around the interplay of health, illness, society, science, and technology. The overarching theme of her research is the ensemble of science, technology, and society, with a focus on nation building and citizenship processes, through the lens of individual bodies and collective practices in Colombia. She has conducted fieldwork in Colombia and in the Netherlands and she works with a network of scholars based in Europe, Latin America, and North America. She works in the Department of Design, Universidad de los Andes, Bogotá.

Eduardo Restrepo is a social anthropologist who has done intensive research on the concept and history of blackness in Colombia, and on constructions of otherness, multiculturalism, and postcolonial theory. His books include Políticas de la teoría y dilemas de los estudios de las colombias negras. He currently works in the Department of Cultural Studies, Pontificia Universidad Javeriana, Bogotá.

Mariana Rios Sandoval has a master's degree in medical anthropology from the University of Amsterdam (2007), which included the thesis "Being a Father, Being a Man: Construction of Masculinity and Fatherhood among Some Men in Mexico City." Before that she received an undergraduate degree in biology at UNAM, Mexico (2003). She is currently working on a master's degree in media studies at the New School, New York (part-time and online), and is research communications coordinator at the Population Council Mexico.

Ricardo Ventura Santos is an anthropologist and senior researcher at the National School of Public Health (Escola Nacional de Saúde Pública) of the Oswaldo Cruz Foundation (Fundação Oswaldo Cruz); he is also associate professor in the Department of Anthropology of the National Museum, Federal University of Rio de Janeiro. He has published extensively on health, demography, race, and science in Brazil. He coauthored the book *The Xavante in Transition: Health, Ecology and Bioanthropology in Central Brazil* (2002) and has coedited *Raça como questão: História, ciência e identidades no Brasil* (2010), "The Biological Anthropology of Living Human Populations: World Histories, National Styles, and International Networks" (special issue of *Current Anthropology*, 2012), and *Racial Identities, Genetic Ancestry, and Health in South America: Argentina, Brazil, Colombia, and Uruguay* (2011).

Ernesto Schwartz-Marín finished his PhD in genomics in society at EGENIS (the ESRC Centre for Genomics in Society) at the University of Exeter in 2012. His PhD dissertation, "Genomic Sovereignty and the Mexican Genome: An Ethnography of Postcolonial Biopolitics," explores the construction of a sovereign realm around human genomic science in Mexico. Previously he completed an MSc in genomics in society at Exeter University and a BA in international affairs at the Instituto Tecnológico de Estudios Superiores de Monterrey. He is currently a research fellow at the University of Durham, developing a comparative project between Mexican and Colombian practices of human identification using forensic genetics and forensic anthropology.

Peter Wade is professor of social anthropology at the University of Manchester and has published widely on issues of racial, ethnic, and national identities in Latin America, particularly Colombia. His books include *Race, Ethnicity and Nation: Perspectives from Kinship and Genetics* (ed., 2007), *Race, Nature and Culture: An Anthropological Perspective* (2002), and *Race and Sex in Latin America* (2009).

index

abstraction of ancestry, 3, 207, 222; domestication and, uncertainties of, 225–26; impersonal abstraction, 220–21

Abu El-Haj, Nadia, 5, 6, 11, 213, 224, 238n1

acculturation, assimilation and, 69–70, 87–88

Acuña-Alonzo, V., Flores-Dorantes, T., Kruit, J. K., et al., 224–25

Acuña Alonzo, Víctor, 217

admixture: admixture mapping, 95, 176, 193, 202; genetic study of, 93–94, 203

African ancestry, genomic proportions of, 215–16

Afrodescendant lineages, 15–16, 77–78, 154, 194, 200–201, 228, 230, 233

Against Race (Gilroy, P.), 49

Agassiz, L., and Cary Agassis, E. C., 33–34, 37, 53

agendas, definition of, 186–94

Agostoni, Claudia, 88

Aguilar Rivera, José Antonio, 85, 89

Aguirre Beltrán, G., and Pozas Arciniega, R., 88

Aguirre Beltrán, Gonzalo, 16, 86, 87

Alberro, Solange, 86

Al cabo de las velas: Expedicione scientíficas en Colombia S. XVIII, XIX y XX (Gómez Gutiérrez, A.), 65

Alcántara, Liliana, 238n4

Alonso, Angela, 143, 147, 152, 153

Alves-Silva, J., da Silva Santos, M., Guimarães, P. E. M., et al., 54n8, 120, 186

América Negra y Oculta (Black and Hidden Americas): Colombia, nation and difference in genetic imagination of, 65–66, 67, 68, 75, 81–82n6, 82–83n15, 82n11

American Journal of Human Biology, 242

American Journal of Human Genetics, 174, 242

American Journal of Physical Anthropology, 179n1, 242

Amerindian ancestry, 6, 53, 115, 122, 149–50, 203–4, 205, 209n16, 225, 233, 234; designation as having, 174–75; genomic proportions of, 215–16

Amerindian lineages, 77, 121, 122–23, 124

ancestries, 2–3, 7; ancestral proportions, 215–16, 217, 218, 221, 223; ancestry informative markers (AIMS), 6, 171, 176–77; biocontinental ancestral contributions, 198; biogeographical ancestries, 2, 5–6, 8, 23, 206, 207, 214, 218, 237; connections of ancestry and disease, uncertainties in, 224–25; genetic reckonings of, 3; Mexican, 86, 91–92, 93, 95, 97, 99, 104, 106n4; Mexican ancestry, demand for samples of, 175; Rio Grande do Sul, heterogeneity of, 120; unlimited finity of ancestry, 222–24

Anderson, Benedict, 132, 161, 221

Andrews, George Reid, 14

Anthias, F., and Yuval-Davis, N., 8

Anthropo-genetic Studies of Isolated Colombian Populations, 61–62

anthropological genetics, 11, 26, 68, 181n15, 247

anthropological stigma, 41

anthropological tradition on moral attribution, 36

anthropology in Mexico, content of, 89–91

Antioqueños in Colombia, 193, 200, 201–2

Antropología Biológica, 66

Anzaldúa, Gloria, 157

Appelbaum, N. P., Macpherson, A. S., and Rosemblatt, K. A., 8, 13, 15, 18, 35, 39, 44

Archivos do Museu Nacional (Brazil), 36

Aréchiga Córdoba, Ernesto, 88

Arellano, J., Vallejo, M., Jimenez, J., et al., 91

Arias, W. H., Rojas, W., Moreno, S., et al., 136

Arocha, Jaime, 66, 192

and, 12–20; racialized categories in Mexico of, 87, 90–91; recentering focus on, 228–32; science and, 21–25, 25–26, 89–92, 169–71, 180nn11–12; stability as assemblage, 191; technological transformation and acceleration of, 69–70; trihybrid model of, 92, 97, 99

mestizo in Mexico: distribution of, 174–75; genomics and, 85–86, 90; historical perspective on, 86–87; idealized image of, 87–88; ideological national icon, 86–89, 104–5; indígena and, relationship between, 161–62, 168, 169–70, 171, 172–73, 174–75, 178; mestizo genomics, objective of, 164; mestizo nation and ancestral components, 186–88; mestizo transformations, 161–78; nation and, 186–94; overlapping with indígenas, 171; power of, 85; recentering the mestizo, 228–32, 238; science and, 89–92, 103–5; underrepresentation of mestizo populations in genomics research, 164; unsettling the mestizo nation, 190–92

methodological nationalism, 25–26

methods, context and: Brazil, genetics research in, 241–43; Colombia, genetics research in, 243–46; Mexico, landscape of human genetics in, 246–48

Mexican Genome Diversity Project (MGDP), 161, 162, 163, 164–65, 166, 167, 168, 169, 171, 172, 173–74, 175–76, 177, 179–80n6, 181nn15–16

Mexico: ancestral contributions of populations, 217; dualistic taxonomy of mixture in, 15–16; landscape of human genetics in, 246–48; social categories and laboratory practices, 185, 187, 188, 192, 194, 198, 199, 203. See also Mexico, possibility of a national genomics in

Mexico, mestizo in biosciences of, 161–78; admixture mapping, 176; Amerindian ancestral population, designation as, 174–75; ancestry informative markers (AIMS), 171, 176–77; biobank creation, 169; biopolitical identity, 161; biopolitical paradigm, 164; biosociality, 177–78; blood sampling, 162–69; cluster building, 172–73; collaboration, ethos of, 177; Computational Genomics Laboratory (INMEGEN), 172; data, mestizo metrics,

172–78; data grouping, 176–77; DNA haplogroup assignment, 169–70; DNA labelling and inferences from, 168–71; donor recruitment, 164–65; EIGENSTRAT technology, 172, 173; European markers, search for, 170–71; genetic differences between Mexican subpopulations, 164; genome-wide association studies, 177; genomic sovereignty, protection of, 175–76; Genotyping and Expression Analysis Unit (INMEGEN), 168, 172; gift relationship, blood donation as, 164–65; HapMap Project, 163–64, 168, 173, 174, 175, 176; health, relationship of blood donation to, 165, 167; homogeneity, ideology of, 164; Human Genome Diversity Project (HGDP), 173; inclusion-and-difference paradigm, 164; indigenous donorship, 163, 165–67, 170, 174, 181n16, 181–82n17; indigenous qualifier, 169, 173; informed consent, 168–69; jornada (work day), 163, 166–67; Maya qualifier, 169; mestizaje, genetic definition of, 170–71; Mexican ancestry, demand for samples of, 175; Mexican Genome Diversity Project (MGDP), 161, 162, 163, 164–65, 166, 167, 168, 169, 171, 172, 173–74, 175–76, 177, 179–80n6, 181nn15–16; "Mexican mestizo population," 161; mitochondrial haplogroups, 169–70; National Institute for Genomic Medicine (INMEGEN), 161–62, 163, 164–65, 166, 167–68, 171, 173, 175, 176, 177, 178, 179n5, 179–80n6, 180n9, 180n13, 181nn15–16, 181–82n17, 182n21; Oaxaca, sampling indigenous populations in, 163, 166, 170, 174, 181n16, 181–82n17; principal component analysis (PCA), 172; recruitmentology, 165–66; sampling crusades, 163–64; sampling sites, delivery of results to, 167–68; single nucleotide polymorphism (SNP) variation scores, 172–73; social relationship between indigenous donors and researchers, 166; subpopulations, recruitment of, 165–66; Supercomputer and Information Technology Unit (INMEGEN), 172; underrepresentation of mestizo populations in genomics research, 164; unified citizenry, 161; validation, process of, 177; Y chromosome haplogroups, 169–70. See also mestizo in Mexico